AIDS Crises

—————— AMONG ——————

African-Americans

—————— CDC reports HIV/AIDS for ——————
African-American Men are 70% of New Infections

DR. ELOISE V. LEWIS

VBerry & Assoc. Inc.

ACKNOWLEDGMENTS

Completing this book would not have been possible without the strength and love of my mother, Leola Francis Berry Donohue, a brilliant woman of courage and faith who always believed in my abilities. Without my mother's faith in me very early in life and beyond, I could not have started or completed this life dream. Thank you to my father, James Lewis I, for financial and moral support all through high school. Special thanks to my cousin Dr. Michele Everett, her brother Raymond Adkins II, and my siblings, all of whom were very supportive during this grueling, yet exciting research process which became, for me, a labor of love. A special expression of aloha to my late maternal grandmother, Gertrude Miriam Green Grobes Berry, second in the line of four generations of women on the Berry branch named Gertrude. Her love and adoration of me are at the pinnacle of my esteem. I would also like to especially acknowledge my mentor, Dr. Anastasia Metros. Your patience and faith in me cannot be measured. Dr. Metros, your brilliant and open mind is a treasure, and I thank you so very much for your talent and patience. I would be remiss not to thank my other committee members, Dr. Louise Underdahl and Dr. Michelle A. Rose, for their dedication to my educational process. Finally, I would like to especially thank two key cohorts, Dr. Lucinda Renee Whitaker and Dr. Kimberly A. McClain, for their dedication and support of my academic efforts. I am forever indebted to each of you.

TABLE OF CONTENTS

LIST OF TABLES

LIST OF FIGURES

1

INTRODUCTION

IN 2008, AN estimated 33.4 million people around the world had human immunodeficiency virus (HIV) and acquired immune deficiency syndrome (AIDS), with 2.7 million newly infected (World Health Organization [WHO], 2006). Both HIV and AIDS have been ravaging the African-American population in the United States. Although African-Americans comprise approximately 13% of the U.S. population, 49% of newly diagnosed cases of HIV in the United States are in African-Americans. African-American heterosexuals (i.e., individuals who have sexual affairs with the opposite sex and without protection, with multiple partners, and in ways that leave skin membranes vulnerable to the virus) are most prone to becoming infected with HIV, followed closely by men who have sex with men (Centers for Disease Control and Prevention [CDC], 2005). The documented prevalence of HIV/AIDS in the African-American U.S. population supported the decision to conduct research and develop proactive strategies to address the problem, with the purpose of identifying potential reasons for such disproportionate numbers.

The existing literature includes potential reasons why the high incidence of HIV/AIDS in African-Americans is not addressed effectively. Faith-based leaders in many predominantly African-American churches may be hesitant to fully acknowledge and effectively address the problem of HIV/AIDS in their congregations. Stigma is a powerful barrier relative to faith-based initiatives in combating HIV/AIDS in the African-American community. The fear of stigma prevents some religious leaders from assuming vocal roles in fighting against HIV/AIDS (Primas, 2008).

There have been at least 11 fairly recent human experiments that took place in America. They were:

The Aversion Project, 1971–1989, involved gay people in apartheid-era South Africa. They didn't like gay people in apartheid-era South Africa, especially in the armed forces. How they got rid of them is shocking. Using army psychiatrists and military chaplains who were, presumably, privy to private, "confidential" confessions, the apartheid regime flushed out homosexuals in the armed forces. But it did not evict them from the military. The homosexual "undesirables" were sent to a military hospital near Pretoria, to a place called Ward 22 (which in itself sounds terrifying).

There, between 1971 and 1989, many victims were submitted to chemical castrations and electric shock treatment, meant to cure them of their homosexual "condition."

As many as 900 homosexuals, mostly 16–24 years old who had been drafted and had not voluntarily joined the military, were subjected to forced "sexual reassignment" surgeries. Men were surgically turned into women against their will, and then cast out into the world, the gender reassignment often incomplete, and without the means to pay for expensive hormones to maintain their new sexual identities (pg. 6).

The Tuskegee Experiments, 1932–1944. There's a good reason many African-Americans are wary of the good intentions of government and the medical establishment. Over the course of the next 40 years, the Tuskegee Study of Untreated Syphilis in the Negro Male denied treatment to 399 syphilitic patients, most of them poor, black, illiterate sharecroppers. Even after penicillin emerged as an effective treatment in 1947, these patients, who were not told they had syphilis, but were informed they suffered from "bad blood," were denied treatment, or given placebo treatments (pg. 3).

The Henrietta Lacks Experiment, 1951. Dr. George Otto Gey of Johns Hopkins University observed the potency of replication of Ms. Lacks' cervical cancer cells and, without her or her family's consent or knowledge, snipped off a portion of her cervix and began harvesting the cells, the first "human immortal" cell line, harvesting over 20 tons of human cells from Lacks' sample.

The Guatemalan STD Study 1940s. Syphilis seemed to bring out the inherent racism in government-funded doctors in the 1940s. Tuskegee's black people weren't the only victims of morally reprehensible studies of this disease. Turns out Guatemalans were also deemed suitable unknowing guinea pigs by the U.S. government.

Penicillin having emerged as a cure for syphilis in 1947, the government decided to see just how effective it was. The way to do this, the government decided, was to turn syphilitic prostitutes loose on Guatemalan prison inmates, mental patients, and soldiers, none of whom consented to be subjects of an experiment. If actual sex didn't infect the subject, then surreptitious inoculation did the trick. Once infected, the victim was given penicillin to see if it worked. Or not given penicillin, just to see what happened, apparently. About a third of the approximately 1,500 victims fell into the latter group. More than 80 "participants" in the experiment died (pg. 10).

Agent Orange Experiments, 1965–1966. Prisoners were injected with dioxin (a toxic byproduct of Agent Orange)—468 times the amount the study originally called for. The results were prisoners with volcanic eruptions of chloracne (severe acne combined with blackheads, cysts, pustules, and other really bad stuff) on the face, armpits, and groin (pg. 12).

Irradiation of Black Cancer Patients, 1960–1971. Exposure of 88 cancer patients, poor and mostly black, to whole-body radiation, even though this sort of treatment had already been pretty well discredited for the types of cancer these patients had. They were not asked to sign consent forms, nor were they told the Pentagon funded the study. They were simply told they would be getting a treatment that might help them. Patients were exposed, in the period of one hour, to the equivalent of about 20,000 x-rays' worth of radiation (pg. 14).

Slave Experiments 1845–1849. It should be no surprise that experiments were often conducted on human chattel during America's shameful slavery history. The man considered the father of modern gynecology, J. Marion Sims, conducted numerous experiments on female slaves between 1845 and 1849. The women, afflicted with vesico-vaginal fistulas, a tear between the vagina and the bladder, suffered greatly from the condition and were incontinent, resulting in societal ostracism (pg. 16).

The Chamber. Back to the Cold War. Prisoners were again the victims, as the Soviet Secret Police conducted poison experiments in Soviet gulags. The Soviets hoped to develop a deadly poison gas that was tasteless and odorless. At the laboratory known as "The Chamber," unknowing and unwilling prisoners were given preparations of mustard gas, ricin, digitoxin, and other concoctions hidden in meals, beverages, or given as "medication" (pg. 18).

World War II: Heyday of Evil Experiments, 1940s. The Tuskegee experiments, the Germans Nazis against the Jews, and the Japanese killed as many as 200,000 people during numerous experimental atrocities in both the Sino-Japanese War and WWII. Some of the experiments put the Nazis to shame (pg. 19).

The Monster Study, 1939. Wendell Johnson, University of Iowa speech pathologist, and his grad student

Mary Tudor conducted stuttering experiments on 22 non-stuttering orphan children. The children were split into two groups. One group was given positive speech therapy, praising them for their fluent speech. The unfortunate other group was given negative therapy, harshly criticizing them for any flaw in their speech abilities and labeling them stutterers (pg. 22).

The result of this cruel experiment was that children in the negative group, while not transforming into full-fledged stutterers, suffered negative psychological effects and several suffered from speech problems for the rest of their lives. Formerly normal children came out of the experiment, dubbed "The Monster Study." –with speech impediments,

Project 4.1, 1952, a medical study conducted on the natives of the Marshall Islands, who in 1952 were exposed to radiation fallout from the Castle Bravo nuclear test at Bikini Atoll, which inadvertently blew upwind to the nearby islands. Instead of informing the residents of the island of their exposure, and treating the victims while they studied them, the U.S. elected instead just to watch quietly and see what happened (pg. 24).

A sordid history of unethical practices such as these has undoubtedly provided a conjuncture for the level of distrust that currently exists between African-Americans, particularly black men, and the medical community.

Trust factors may also have a bearing on African-Americans' openness to treatment and prevention efforts.

Lack of trust may be associated with past events in African-American history, such as the Tuskegee experiments conducted in Tuskegee, Alabama, between 1932 and 1972 by a group of doctors and other health professionals to determine the outcomes of untreated syphilis. The experiments were conducted on 399 African-American men, without significant attention to ethical practices (Thomas & Quinn, 1991).

The Tuskegee experiments were dubbed an American travesty and were infamous in the medical research community. Ultimately they were recognized as abusive, racist, bureaucratically connected, and driven by personal motives. The medical personnel involved in the Tuskegee experiments remained unmonitored and uncensored during 40 years of abuse. The research participants were intentionally harmed and were uncompensated during the experiments, in large part because of a lack of oversight from medical research review boards and loose regulations for the protection of research participants (Lombardo & Dorr, 2006).

The Presidential Commission for the Study of Bioethical Issues imposed regulations to protect and compensate research participants falling victim to unethical research methods (Basken, 2011). It was not until December 2011, 37 years after the conclusion of the Tuskegee experiments, that the Presidential Commission for the Study of Bioethical Issues imposed regulations to protect and compensate research participants falling victim to unethical research methods (Basken, 2011). The regulations were established shortly after American scientists conducted another unethical study with Guatemalans. The commission was formed in November, 2010 after the revelation that U.S. scientists intentionally infected hundreds of people in Guatemala with gonorrhea and syphilis as part of a public health research project. The U.S. president at the time formed the commission to determine whether existing guidelines were sufficient to protect human participants. The commission reviewed the Guatemala case along with nine others, including an AIDS study in Africa in the 1990s in which pregnant women were given placebos, even though the drug AZT was known to be effective in saving the lives of children (Basken, 2011).

The Tuskegee experiments and other ethically questionable practices and experiments may have added to the African-American experience of abuse, starting with the slavery experience of over 200 years. Combined, these events may have set the stage for a metaphorical wall of distrust between African-Americans and the American medical community. Extensive research is necessary to (a) understand African-Americans' perceptions of trust in the American medical system and (b) introduce the possibility that a lack of trust resulting from such events as the Tuskegee experiments contributes to behaviors that increase the prevalence of HIV/AIDS in African-Americans (Thomas & Quinn, 1991).

The intent of the current study was to better understand African-Americans' perceptions of past unethical research in relation to their sexual health choices and HIV/AIDS. The terms "African-Americans," "black people," and "blacks" are used interchangeably. The epidemiology of HIV/AIDS among African-Americans has remained consistently higher than other races for more than a decade. Despite growing efforts to combat HIV/AIDS in America, African-Americans continue to represent approximately 44% of new cases, while comprising only approximately 14% of the population in the United States of America (USA). This hermeneutic phenomenological research study explored African-Americans' perceptions of the relationship between genocidal conspiracy beliefs and sexual health behavior through the lived experiences of 16 African-American men and women living in the United States, to better understand whether those perceptions might influence behaviors that can contribute to higher incidence of HIV/AIDS in the African-American community. The five overarching themes emerging from the data concluded: (a) belief that a virus was created; (b) homosexual behavior as an implication; (c) lack of access to health care as a threat; (d) lack of education about HIV/AIDS; and (e) stigmatism, ostracism, shame, shunning, and stereotyping of persons diagnosed with HIV, all as one common theme. Leadership implications highlighted the need for nationwide screening and increased efforts to educate all Americans on HIV/AIDS, since one in five Americans does not know they are HIV-positive.

The epidemiology of HIV/AIDS among African-Americans has remained consistently higher than other races for more than a decade.

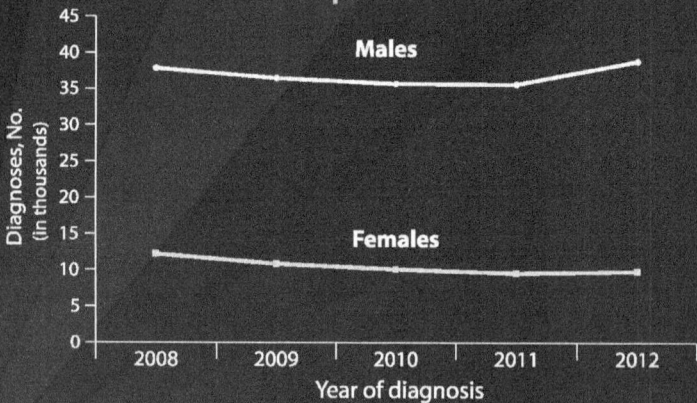

CDC (2012)

Diagnoses of HIV Infection among Adults and Adolescents, by Transmission Category, 2008–2012—United States and 6 Dependent Areas

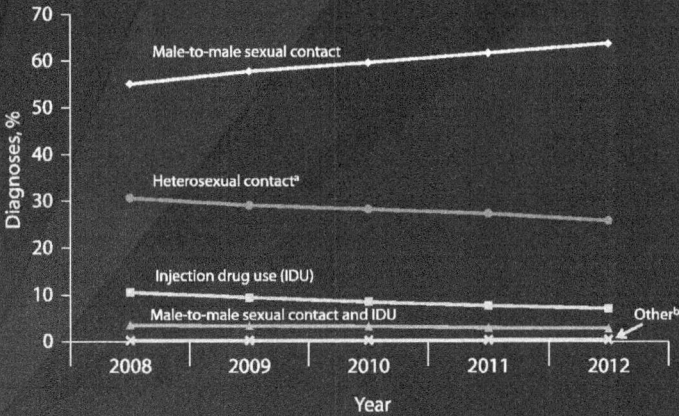

Note. Data include persons with a diagnosis of HIV infection regardless of stage of disease at diagnosis. All displayed data have been statistically adjusted to account for reporting delays and missing transmission category, but not for incomplete reporting.
[a] Heterosexual contact with a person known to have, or to be at high risk for, HIV infection.
[b] Includes hemophilia, blood transfusion, perinatal exposure, and risk factor not reported or not identified.

Diagnoses of HIV Infection among Adults and Adolescents, by Sex and Transmission Category, 2012—United States and 6 Dependent Areas

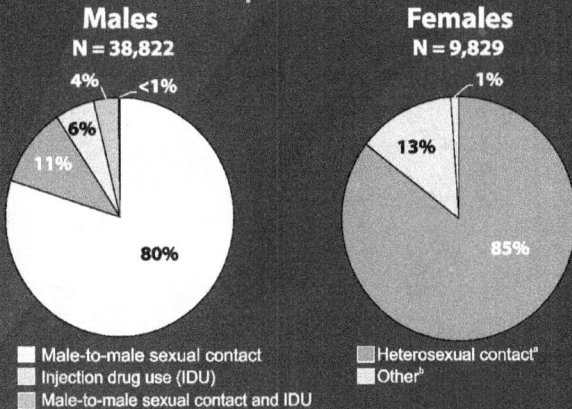

Note. Data include persons with a diagnosis of HIV infection regardless of stage of disease at diagnosis. All displayed data have been statistically adjusted to account for reporting delays and missing transmission category, but not for incomplete reporting.
[a] Heterosexual contact with a person known to have, or to be at high risk for, HIV infection.
[b] Includes hemophilia, blood transfusion, perinatal exposure, and risk factor not reported or not identified.

CDC (2012)

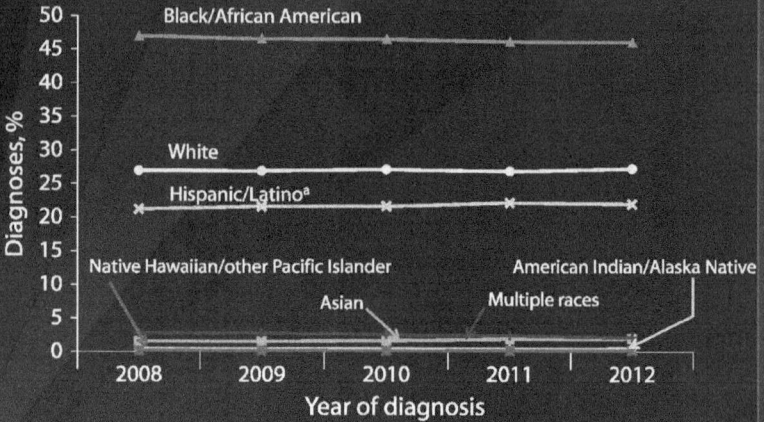

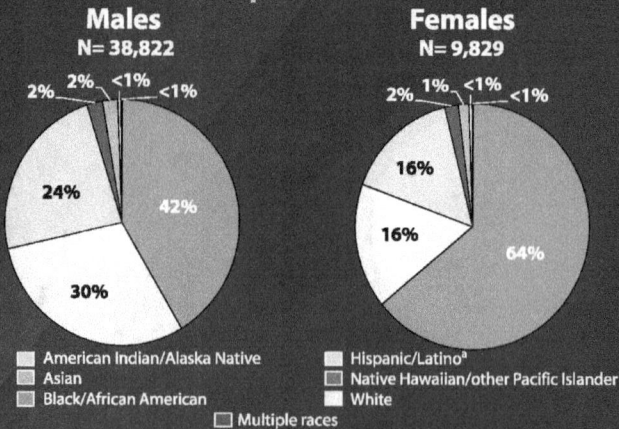

CDC (2012)

Diagnoses of HIV Infection among Adults and Adolescents, by Age at Diagnosis, 2012—United States

N = 47,746

- 13–24 years
- 25–34 years
- 35–44 years
- 45–54 years
- ≥55 years

9% · 19% · 29% · 21% · 22%

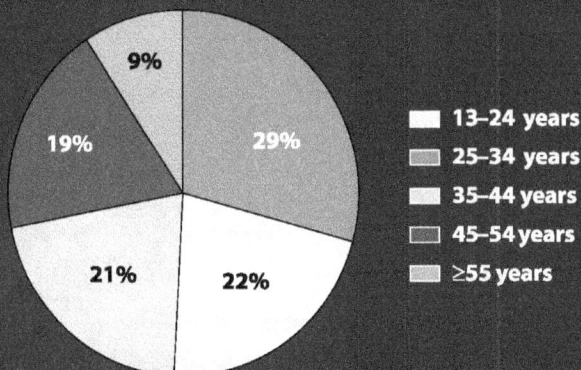

Note. Data include persons with a diagnosis of HIV infection regardless of stage of disease at diagnosis. All displayed data have been statistically adjusted to account for reporting delays, but not for incomplete reporting.

Diagnoses of HIV Infection among Adults and Adolescents, by Transmission Category, 2012—United States and 6 Dependent Areas

Transmission Category	No.	%
Male-to-male sexual contact	31,049	63.8
Injection drug use (IDU)	3,456	7.1
Male-to-male sexual contact and IDU	1,375	2.8
Heterosexual contact[a]	12,580	25.9
Other[b]	191	0.4
Total[c]	48,651	100.0

Note. Data include persons with a diagnosis of HIV infection regardless of stage of disease at diagnosis. All displayed data have been statistically adjusted to account for reporting delays and missing transmission category, but not for incomplete reporting.
[a] Heterosexual contact with a person known to have, or to be at high risk for, HIV infection.
[b] Includes hemophilia, blood transfusion, perinatal exposure, and risk factor not reported or not identified.
[c] Because column totals for estimated numbers were calculated independently of the values for the subpopulations, the values in each column may not sum to the column total.

CDC

CDC (2012)

Diagnosed HIV Infections Attributed to Male-to-Male Sexual Contact, by Race/Ethnicity, 2012—United States and 6 Dependent Areas

Race/ethnicity	No.	%
American Indian/Alaska Native	132	0.4
Asian	710	2.3
Black/African American	11,959	38.5
Hispanic/Latino[a]	7,405	23.9
Native Hawaiian/other Pacific Islander	59	0.2
White	10,072	32.4
Multiple races	711	2.3
Total[b]	31,049	100.0

Note. Data include persons with a diagnosis of HIV infection regardless of stage of disease at diagnosis. All displayed data have been statistically adjusted to account for reporting delays and missing transmission category, but not for incomplete reporting.
[a] Hispanics/Latinos can be of any race.
[b] Because column totals for estimated numbers were calculated independently of the values for the subpopulations, the values in each column may not sum to the column total.

CDC

Diagnosed HIV Infections Attributed to Heterosexual Contact[a], by Sex and Race/Ethnicity, 2012—United States and 6 Dependent Areas

Race/ethnicity	Males No.	Males %	Females No.	Females %	Total No.	Total %
American Indian/Alaska Native	16	0.4	32	0.4	48	0.4
Asian	48	1.1	135	1.6	183	1.5
Black/African American	2,745	65.7	5,542	66.0	8,288	65.9
Hispanic/Latino[b]	801	19.2	1,315	15.7	2,117	16.8
Native Hawaiian/other Pacific Islander	2	0.0	9	0.1	10	0.1
White	508	12.2	1,225	14.6	1,733	13.8
Multiple races	57	1.4	144	1.7	201	1.6
Total[c]	4,177	100.0	8,402	100.0	12,580	100.0

Note. Data include persons with a diagnosis of HIV infection regardless of stage of disease at diagnosis. All displayed data have been statistically adjusted to account for reporting delays and missing transmission category, but not for incomplete reporting.
[a] Heterosexual contact with a person known to have, or to be at high risk for, HIV infection.
[b] Hispanics/Latinos can be of any race.
[c] Because column totals for estimated numbers were calculated independently of the values for the subpopulations, the values in each column may not sum to the column total.

CDC

CDC (2012)

Rates of Diagnoses of HIV Infection among Adults and Adolescents, 2012—United States and 6 Dependent Areas

N = 48,651   Total Rate = 18.4

NH	4.7
MA	18.1
RI	9.2
CT	10.5
NJ	24.4
DE	19.4
MD	36.6
DC	160.7

Rates per 100,000 population
- <10.0
- 10.0 – 19.9
- 20.0 – 29.9
- ≥30.0

American Samoa	0.0
Guam	2.1
Northern Mariana Islands	0.0
Puerto Rico	28.8
Republic of Palau	0.0
U.S. Virgin Islands	17.1

Note. Data include persons with a diagnosis of HIV infection regardless of stage of disease at diagnosis. All displayed data have been statistically adjusted to account for reporting delays, but not for incomplete reporting.

CDC

Rates of Adults and Adolescents Living with Diagnosed HIV Infection, Year-end 2011—United States and 6 Dependent Areas

N = 896,621   Total Rate = 342.1

NH	99.8
MA	323.0
RI	219.2
CT	331.2
NJ	492.5
DE	381.1
MD	588.9
DC	2,721.6

Rates per 100,000 population
- <100.0
- 100.0 – 199.9
- 200.0 – 299.9
- 300.0 – 399.9
- ≥400.0

American Samoa	2.4
Guam	68.9
Northern Mariana Islands	6.0
Puerto Rico	585.0
Republic of Palau	24.2
U.S. Virgin Islands	685.1

Note. Data include persons with a diagnosis of HIV infection regardless of stage of disease at diagnosis. All displayed data have been statistically adjusted to account for reporting delays, but not for incomplete reporting.

CDC

CDC (2012)

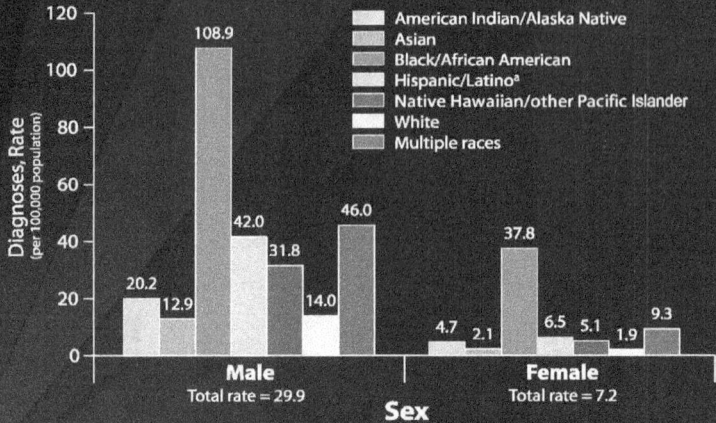

Rates of Diagnoses of HIV Infection among Adults and Adolescents, by Sex and Race/Ethnicity, 2012—United States

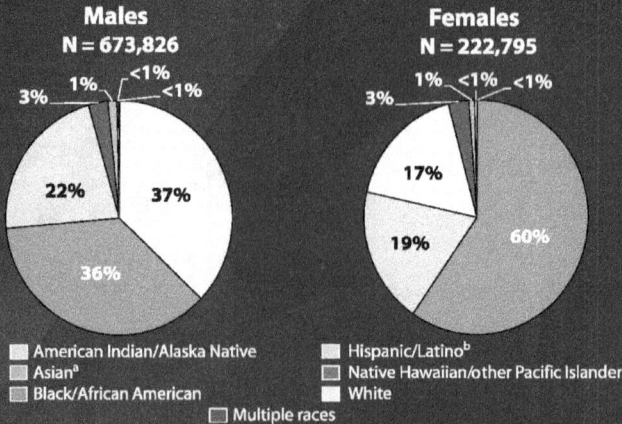

Adults and Adolescents Living with Diagnosed HIV Infection, by Sex and Race/Ethnicity, Year-end 2011— United States and 6 Dependent Areas

CDC (2012)

Adults and Adolescents Living with Diagnosed HIV Infection, by Sex and Transmission Category, Year-end 2011— United States and 6 Dependent Areas

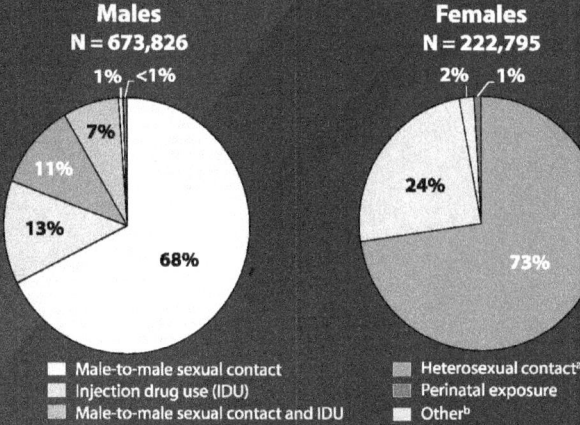

Males N = 673,826

Females N = 222,795

Males: 68%, 13%, 11%, 7%, 1%, <1%

Females: 73%, 24%, 2%, 1%

- Male-to-male sexual contact
- Injection drug use (IDU)
- Male-to-male sexual contact and IDU
- Heterosexual contact[a]
- Perinatal exposure
- Other[b]

Note. Data include persons with a diagnosis of HIV infection regardless of stage of disease at diagnosis. All displayed data have been statistically adjusted to account for reporting delays and missing transmission category, but not for incomplete reporting.
[a] Heterosexual contact with a person known to have, or to be at high risk for, HIV infection.
[b] Includes hemophilia, blood transfusion, and risk factor not reported or not identified.

Percentages of Stage 3 (AIDS) Classifications among Adults and Adolescents with HIV Infection, by Race/Ethnicity and Year of Diagnosis, 1985–2012—United States and 6 Dependent Areas

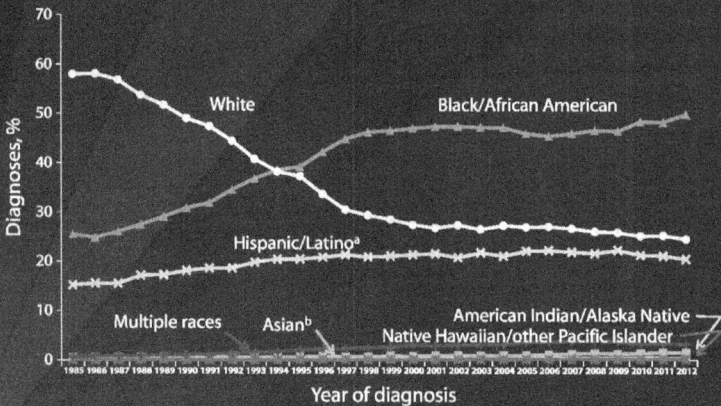

White
Black/African American
Hispanic/Latino[a]
Multiple races   Asian[b]
American Indian/Alaska Native
Native Hawaiian/other Pacific Islander

Year of diagnosis

Note. All displayed data have been statistically adjusted to account for reporting delays, but not for incomplete reporting.
[a] Hispanics/Latinos can be of any race.
[b] Includes Asian/Pacific Islander legacy cases.

CDC (2012)

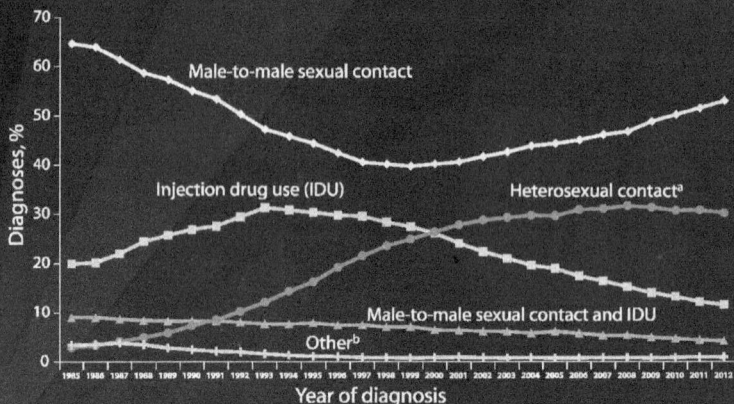

Percentages of Stage 3 (AIDS) Classifications among Adults and Adolescents with HIV Infection, by Transmission Category and Year of Diagnosis, 1985–2012—United States and 6 Dependent Areas

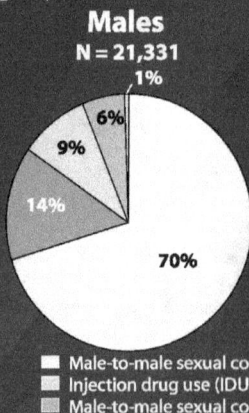

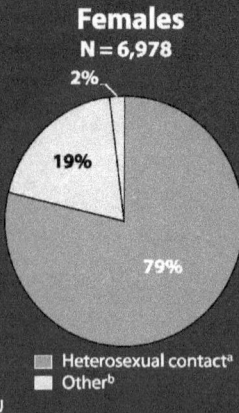

Stage 3 (AIDS) Classifications among Adults and Adolescents with HIV Infection, by Sex and Transmission Category, 2012—United States and 6 Dependent Areas

CDC (2012)

Stage 3 (AIDS) Classifications among Adults and Adolescents with HIV Infection, by Race/Ethnicity, 2012—United States

Race/ethnicity	No.	Rate
American Indian/Alaska Native	114	6.1
Asian[a]	418	3.2
Black/African American	14,094	44.8
Hispanic/Latino[b]	5,418	13.5
Native Hawaiian/other Pacific Islander	32	7.7
White	6,932	4.1
Multiple races	910	23.8
Total[c]	**27,918**	**10.7**

Note. All displayed data have been statistically adjusted to account for reporting delays, but not for incomplete reporting.
Rates are per 100,000 population.
[a] Includes Asian/Pacific Islander legacy cases.
[b] Hispanics/Latinos can be of any race.
[c] Because column totals for estimated numbers were calculated independently of the values for the subpopulations, the values in each column may not sum to the column total.

Adults and Adolescents Living with Diagnosed HIV Infection Ever Classified as Stage 3 (AIDS), by Sex, 1993–2011—United States and 6 Dependent Areas

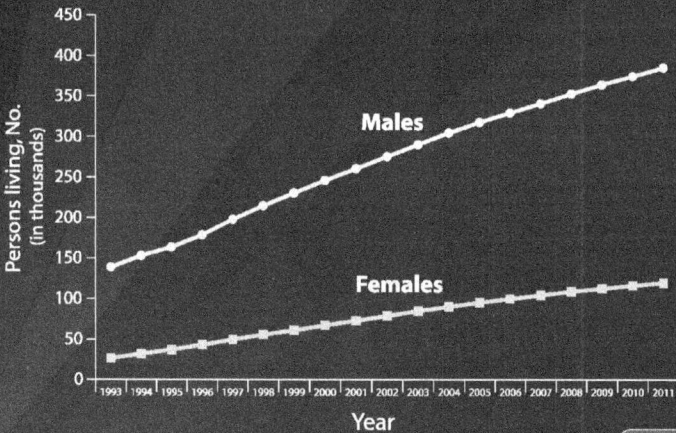

Note. All displayed data have been statistically adjusted to account for reporting delays, but not for incomplete reporting.

CDC (2012)

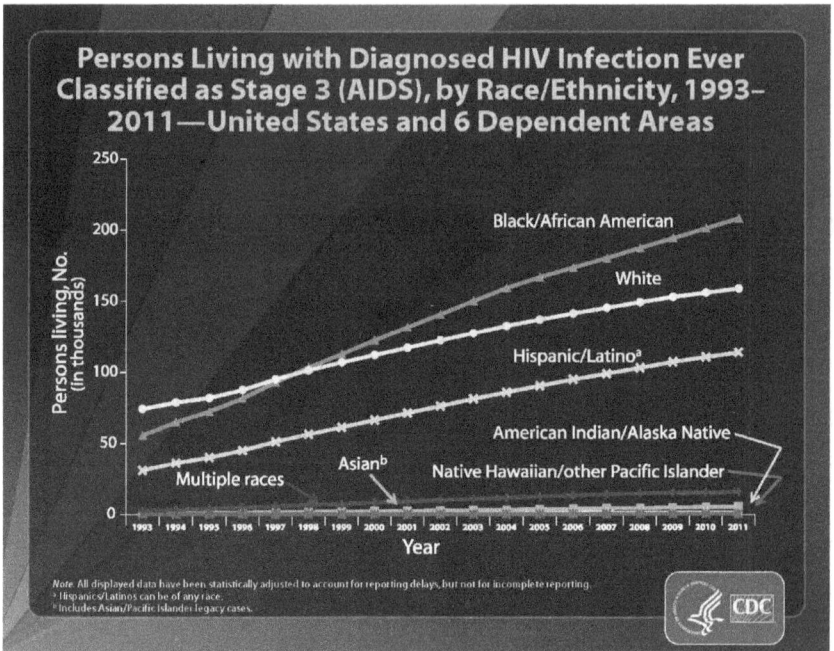

Persons Living with Diagnosed HIV Infection Ever Classified as Stage 3 (AIDS), by Race/Ethnicity, 1993–2011—United States and 6 Dependent Areas

CDC (2012)

Background of the Problem

DEFINING HIV

The Human Immunodeficiency Virus (HIV) is defined as a positive result on a screening test for HIV antibody, followed by a positive result on a confirmatory test for HIV antibody (CDC, 2008). The definition of AIDS has been revised several times over the course of the past two decades (CDC, 2007a), and is explained in detail within the classification tables presented in a later chapter. The first cases of HIV/AIDS in the United States were documented in the late 1980s in the gay male population (CDC, 2007a). Transmission of HIV/AIDS occurs in a variety of ways, such as using hypodermic needles previously used by infected individuals, receiving unfiltered blood transfusions, and engaging in high-risk sexual behaviors (Ezejiofor, 2008). High-risk sexual behaviors include having multiple sexual partners, unprotected oral and anal sex, and sexual intercourse without protection during menses, as well as men engaging in sexual activities with other men (known as *fisting*).

RISKY BEHAVIOR

Engaging in compromising intimate behavior increases the odds of contracting and spreading HIV/AIDS. According to the CDC (2009), a risky sexual behavior involves persons having sex with uninfected people without divulging their HIV/AIDS status. Sometimes infected individuals do not know they are infected and unknowingly transmit the infection to sexual partners during unprotected sex. Deep kissing can lead to contracting HIV (Ezejiofor, 2008).

The disparity in HIV/AIDS infection is particularly apparent for African-Americans, who constitute only 13% of the U.S. population but make up 49% of all HIV/AIDS patients. Approximately 500,000 African-Americans are living with HIV/AIDS (Wyatt, Williams, Henderson, & Sumner, 2009). Primas (2008) noted a shift in the prevalence of HIV/AIDS infection from mostly gay Caucasian men to African-Americans, with African-American women accounting for 67% of all new cases among African-Americans. The pervasiveness of HIV/AIDS among African-American women poses a major health risk, because when infected women become pregnant, they can pass the infection on to their babies.

Wyatt et al (2009) noted multiple reasons for the high incidence of HIV/AIDS in the African-American population, citing behaviors that increase risk, such as poverty, chronic homelessness, prison time, and limited access to health care. According to a CDC (2006) report, other risk factors include (a) unprotected sexual practices among high-risk groups in the African-American culture, particularly among men who have sex with other men; (b) alcohol consumption; (c) intravenous drug users sharing needles;

and (d) heterosexual individuals having unprotected sex with others who are oblivious to their HIV health status. These factors may contribute to the disproportionate incidence of HIV/AIDS in the African-American community (CDC, 2006), but reasons why the epidemiology for HIV/AIDS in the African-American community is higher than in the rest of the population remain unclear.

INSTITUTIONAL MISTRUST

Other factors may influence African-Americans' sexual health decisions and increase their likelihood of becoming infected with HIV/AIDS. One factor may be distrust of the medical community resulting from events such as the Tuskegee experiments. Mistrust and conspiratorial beliefs regarding the medical community could play a role in the sexual behavior choices of African-Americans, including the decision of whether to seek screening, vaccinations, and medical care for HIV/AIDS. Whether the physicians involved in the Tuskegee experiments lacked ethical integrity based on conscious intention remains open to question. Various books, journals, and other credible sources include accounts of scientists involved in the Tuskegee experiments consciously neglecting to provide the full spectrum of information to participants, including not revealing the extent to which participants would be exposed to the syphilis virus (Thomas & Quinn, 1991).

HISTORY OF QUESTIONABLE ETHICS

In addition to the distrust the Tuskegee experiments potentially generated, the history of African-American slavery established a relationship between African-Americans and Caucasians in which Caucasians held significant power and control (Gilley & Keesee, 2007). According to The Peculiar Institution (2009), in the 1857 ruling regarding Dred Scott's appeal for freedom, members of the Supreme Court determined that black people could never be citizens of the United States. The Missouri Compromise was ruled unconstitutional, thereby allowing slavery in all U.S. territories. A parallel may exist between the slave experience in America and ideations of conspiracy related to the AIDS crisis. In regard to American Indians' experience, Gilley and Keesee (2007) stated,

> Conspiracy beliefs, we argue, should be looked at as a potential form of power recognition, where American Indians draw on their experiences of oppression to explain the presence of HIV/AIDS within their communities, at the same time that they draw on public health knowledge to explain how humans get HIV/AIDS. (p. 44)

Another well-publicized research experiment stretched the limits of ethics. On February 8, 1951,

American physicians and scientists used 30-year-old African-American Henrietta Lacks' cervical tissue and cell lining for scientific research (Lucey, Nelson-Rees, & Hutchins, 2009). The cancer cells, named HeLa cells, were found to be extremely beneficial for future success in the treatment of cancer. Several other areas of Lacks' anatomy were not examined because permission was not granted, but the cancerous cells were taken from her cervix without permission because the material was deemed discarded material, needing no consent (Lucey et al, 2009).

Henrietta Lacks' cervical cells contributed to a scientific breakthrough because of the discovery of the HeLa cells' exponential growth factor, supporting therapy for new vaccines. Lacks' cervical biopsy supplied tissue to the pathology department for clinical evaluation and to the Johns Hopkins Hospital for research purposes (Lucey et al, 2009). The two significant figures making important contributions to molecular biology on behalf of Henrietta Lacks' HeLa cervical cells were Ross Harrison, MD, PhD, and Warren Lewis, MD.

No compensation or recognition was ever extended to Lacks or her family for her cervical cells (Lucey et al, 2009). This event occurred approximately ten years after the Tuskegee experiments. Such examples of unethical medical experimentation on African-Americans support the theory that a link might exist between histories of oppression and beliefs of conspiracy and power perpetrated against minorities in America, setting the stage for possible exploitation for scientific gain.

The intent of the present hermeneutic, phenomenological research study was to obtain the perspectives of 18- to 65-year-old African-Americans about conspiracy theories and sexual health choices. Understanding the participants' perspectives may lead to the implementation of interventions that will greatly decrease the number of African-Americans who become infected with HIV/AIDS. The phenomenological design facilitated obtaining a deep understanding of the participants' experiences and perspectives. The phenomenological design usually involves conducting individual interviews, an appropriate data collection method to capture the lived experiences of participants and gain insight about a phenomenon of interest (Creswell, 2008).

Problem Statement

The general problem addressed in the study is that African-Americans continue to be the highest-risk group for HIV/AIDS infection and may consequently infect others, knowingly or unknowingly (Primas, 2008). Mistrust of the American medical community and deep-seated and subconscious ideas of conspiracy could be hampering efforts to treat and prevent HIV/AIDS. Erroneous perceptions on how HIV is transmitted may also contribute to the high incidence of HIV/AIDS.

A scene in the movie *Precious,* which depicts a mother who contracted HIV from her husband, provides an example of this phenomenon. When consulting with a social worker about her recently acquired HIV status, the mother explains she did not screen for HIV sooner because she was under the impression that the disease was spread through anal sex only (Winfrey, Perry, & Daniels, 2009). Misunderstandings about HIV/AIDS indicate that public service messages and other efforts to educate African-Americans about transmission of HIV are not completely successful.

The specific problem addressed in the present research is that African-Americans' distrust of the medical community may negatively influence their sexual health choices (e.g., not requesting screening for HIV/AIDS). There is little research on the potential influence of distrust and conspiracy theories on African-Americans' sexual health choices and likelihood of contracting HIV/AIDS. The purpose of this hermeneutic, phenomenological research study was to explore the lived experiences of African-American men and women and obtain their perspectives on genocide conspiracy and sexual health practices in relation to the high rates of HIV/AIDS among blacks. African-Americans between 18 and 65 years old participated in interviews and shared in-depth information about the phenomenon under study.

Purpose Statement

The purpose of this hermeneutic, phenomenological research study was to explore the perceptions of African-American men and women of genocide conspiracy and sexual health practices in relation to high rates of HIV/AIDS among blacks. The qualitative investigative approach involved conducting one-on-one interviews to collect narrative data on the participants' experiences and perceptions. Data were collected from male and female African-Americans, age 18 to 65, living in various states, including Hawaii.

This population was chosen to fulfill one of the objectives of the study because of its highest epidemiology of HIV/AIDS (CDC, 2009). Interviewing members of the study population resulted in collecting firsthand, in-depth data regarding the experiences and perceptions of African-Americans. Some of the participants may have been infected with HIV/AIDS or have family members, friends, or associates infected with HIV/AIDS.

The qualitative approach and phenomenological method involve investigating deep, conscious thoughts about experiences and perceptions, such as the perception of mistrust of the American medical community. The basis of qualitative research is inductive reasoning that leads to a deep understanding of a phenomenon. In the present study, inductive reasoning helped identify possible reasons for the consistently high rates of HIV/AIDS among African-Americans.

The phenomenological method was appropriate to explore the understanding of people's ontological worldview, perceptions, ideals, and perspectives of a scenario. The goal of phenomenology is to glean from participants descriptions of what it is like to experience a particular situation that is real to them (Leedy & Ormrod, 2005). Phenomenology is used to examine critical situations deemed societal problems and make sense of the chaos.

The phenomenological method included face-to-face, in-depth interviews, bracketing and coding, and tape-recorded telephone interviews. The data provided an in-depth understanding of the perspectives and experiences of African-Americans in relation to conspiracy percep-tions and the disease and epidemiology of HIV/AIDS in the African-American community. Bracketing and cod-ing are qualitative methodology tools needed to complete the process of (a) identifying repeated themes that emerge from field notes and interviews and (b) extracting over-arching themes that address the central research question of the study (Moustakas, 1994; Neuman, 2003).

Significance of the Study

This qualitative, hermeneutic, phenomenological re-search study has significance for several reasons. For almost a decade, the epidemiology of HIV/AIDS among African-Americans has remained consistently higher than among oth-er racial groups (CDC, 2007b) despite various community outreach efforts to screen African-Americans for, and educate minorities on, HIV/AIDS. Smallman (2008) argued that the continued epidemiological spread of HIV/AIDS can, over time, threaten the entire African-American population in the United States if the rising prevalence of HIV/AIDS among this population is not contained.

Educational messages about the transmission of HIV and AIDS are not decreasing the incidence of HIV/AIDS cases to the degree expected. African-Americans have differing perceptions on how the disease is transmitted, and do not always take precautionary measures. Ongoing notions of conspiracy are lingering, which may over-shadow the efforts to prevent the spread of HIV/AIDS (Winfrey et al, 2009). The findings of this hermeneu-tic, phenomenological research study included details of African-Americans' perceptions regarding the Tuskegee experiments, the genocide conspiracy, and sexual health decisions in relation to HIV/AIDS.

The findings of the present study added to the ex-isting body of research and can be used by government leaders to address the high rates of HIV/AIDS in the African-American population. The findings are particu-larly meaningful in terms of the *triple-loop learning pro-cess.* Triple-loop learning consists of gaining insight into the nature of a paradigm, as opposed to merely assessing which paradigm is superior to another (Peschl, 2007).

Triple-loop learning leads to abundant insight that may otherwise be missed or deemed invaluable, and promotes creative and unorthodox solutions to a problem. An illustration of triple-loop learning occurs in the film *Apollo 13*, with the line, "Houston, we have a problem" (Grazer & Howard, 1995, pg. 1). When the spaceship encounters serious engine trouble, Tom Hanks taps into the intelligence of the crew through dialogue that leads to triple-loop learning. With this process, the crew creates the technology needed to operate the spaceship without traditional resources.

The dialogue in the current study led to triple-loop learning regarding African-Americans' perceptions and experiences of the phenomenon under study. Leaders might use the findings of the inquiry to create and implement new strategies to educate African-Americans on HIV/AIDS. Such methods may decrease the prevalence of HIV/AIDS among African-Americans, which would help protect the African-American population.

Nature of the Study

The purpose of this hermeneutic, phenomenological research study was to explore the lived experiences of African-American men and women and obtain their perspectives of genocide conspiracy and sexual health practices in relation to high rates of HIV/AIDS among blacks in America. The qualitative method and the phenomenological design were the most appropriate choices among available methodologies to fulfill the purpose of the study.

Other qualitative methods such as case study, ethnography, grounded theory, and historical research would not have been appropriate. A quantitative or a mixed-research design involving a combination of quantitative and qualitative data would not have served the purpose outlined for the research. The goal of the study was to understand the lived experiences and perceptions of African-Americans regarding the phenomenon of the high incidence of HIV/AIDS among the African-American population. The most appropriate method was the hermeneutic, phenomenological method that is based on the exploration of participants' ontologies. One-on-one interviews facilitated the access to the participants' experience on deeper levels than could have been captured in a survey or quantitative questionnaire.

Qualitative research is a holistic approach to information that generates a wide spectrum of facets in the phenomenon under study. The qualitative method is a valid and valuable method of research that involves (a) using a philosophical research statement to guide the study; (b) identifying a sound theoretical framework; (c) documenting the research methodology; (d) evaluating qualitative data based on scientific criteria; and (e) explaining the findings in a manner that enhances the credibility, confirmability, dependability, and transferability of findings (Ary, Jacobs, & Sorensen, 2006). The nature of phenomenological, hermeneutic, qualitative research was aligned with the purpose and scope of the present research study, making the phenomenological methodology the most appropriate choice.

The mixed method that incorporates aspects of qualitative and quantitative research was considered but not chosen, because the focus of the study was on obtaining in-depth textual data from one-on-one interviews, not a mix of textual and numerical data (Ary et al, 2006). Qualitative research generates broader information than quantitative research. In qualitative interviews, research participants draw vivid pictures that describe a problem or phenomenon. The qualitative method was preferred over the quantitative method.

Of the various qualitative research designs, the phenomenological design offered the most appropriate methodology for the study. Case studies are appropriate when the purpose is to examine one individual, group, organization, or program to obtain an overall perspective of the issue being studied. Grounded theory involves analyzing data to develop a theory (Ary et al, 2006). Using the phenomenological method resulted in exploration of the perceptions and experiences of participants and their sexual health practices, the Tuskegee experiments, and the genocide conspiracy in relation to HIV/AIDS. This exploration resulted in a better understanding of factors related to the prevalence of HIV/AIDS among the African-American population.

Research Questions

The purpose of this hermeneutic, phenomenological research study was to explore the lived experiences of African-American men and women and obtain their perspectives of genocide conspiracy and sexual health practices in relation to high rates of HIV/AIDS among blacks in America. The following research questions helped accomplish the goal of the study:

R1.   What are the lived experiences and perceptions of African-Americans pertaining to conspiracy beliefs and the influence of such beliefs on their sexual behaviors?

R2.   What influence, if any, do conspiracy beliefs held by African-Americans have on their decisions associated with seeking medical attention?

A qualitative, phenomenological approach facilitated obtaining data with open-ended questions during face-to-face and telephone interviews with participants who gave their consent. The data obtained provided answers to the research questions guiding the study. Qualitative methodology involves a variety of constructs, and scholars such as Creswell (2008) and Denzin and Lincoln (2000) provided ample appropriate techniques for gathering and analyzing qualitative data.

Theoretical Framework

A transformational leadership paradigm to explain current behaviors and predict future behaviors served as theoretical framework for this qualitative, hermeneutic research study. The study topic was HIV/AIDS epidemiology in the African-American community. The selected theoretical framework provided a foundation for exploring the lived experiences of African-American men and women and the influence on their sexual behaviors of their perceptions on genocide conspiracy theories.

Bogart and Bird (2005) conducted culturally focused research and concluded, "HIV/AIDS conspiracy beliefs are a barrier to HIV prevention among African-Americans and may represent a facet of negative attitudes about condoms among Black men" (pg. 218). In peer-reviewed journal literature, Bogart and Bird supported the need to develop trust within the African-American community where issues of mistrust exist. Mistrust is associated with beliefs in conspiracies to use African-Americans as guinea pigs for scientific research. Male participants, in particular, tend to believe that activities conducted to harm African-Americans were the result of a conspiracy.

The theoretical framework for the study included shame and ostracism as factors or barriers to sexual health. The participants shared their perceptions and experiences for exploration with the goal of understanding why the incidence of HIV/AIDS is consistently higher among

African-Americans than among other racial groups. Aggressive efforts to combat the disease through the influences of the media, government, religious leaders, educational leaders, and medical practitioners have not reduced the incidence of the disease. The prevalence of HIV/AIDS among African-Americans has grown steadily since the beginning of the 21st century.

Though the gap is narrowing between African-Americans and other racial groups for HIV/AIDS rates (WHO, 2006), more information is needed to understand the consistently higher epidemiology of HIV/AIDS among African-Americans and effectively combat HIV/AIDS within the black community in the United States. Most efforts to fight HIV/AIDS have focused on Africa.

Several countries in sub-Saharan regions have received 17 *Secure the Future* grants, totaling $31.5 million, to increase efforts for research and development, education, and community outreach projects for HIV/AIDS. The United States received $7.5 million. "Secure the Future is the U.S. $100 million commitment by Bristol-Myers Squibb Co. to assist women and children infected and affected with HIV/AIDS in Botswana, Lesotho, Namibia, South Africa, and Swaziland" (Bristol-Myers Squibb Co., 2000, pg. 1). Despite the overwhelming monetary support to various countries in Africa to battle HIV/AIDS, the epidemiology of the disease remains high.

Definitions of Terms

The following definitions of ten key terms associated with the context and content of the study help define the specific use of the most salient terms:

AIDS (acquired immune deficiency syndrome). According to Schneider et al (2008), *"All HIV-infected persons with a CD4+ T-lymphocyte count of < 200 cells/μL or a CD4+ T- lymphocyte percentage of total lymphocytes of < 14 and 2)"* (pg. 2).

Communication. Communication is "a process by which information is exchanged between individuals, through a common system of symbols, signs, or behavior" ("Communication," 2007, pg. 3).

Communication deviance (CD). Communication deviance is "an inability for a family to share a focus of attention" (Mikesell, Lusterman, & McDaniel, 2003, pg. 184).

Communication failure. "Failure to communicate is one of the most important immediate causes of sexual disorders. Many people expect their partners to have ESP concerning their own sexual needs" (Hyde & DeLamater, 2003, pg. 502).

Conspiracy theories. Fenster (2008) explained, "Conspiracy theories occupy an important place in American democracy. Conspiracy theories are recurrent, associated with major historical events, and circulated through mass culture and politics" (pg. 1).

HIV (human immunodeficiency virus). HIV is "a positive result on a screening test for HIV antibody...followed by a positive result on a confirmatory test for HIV antibody (e.g., Western Blot or immunofluorescence antibody or test)" (CDC, 2007a, pg. 1).

Phenomenological approach. The phenomenological approach to research is "a philosophy of research that focuses on understanding the meaning events have for people in particular situations" (Ary et al, 2006, pg. 647).

Phenomenological interviewing. Phenomenological interviewing consists of "examining lived experience through a series of in-depth interviews" (Ary et al, 2006, pg. 647).

Stigma. A stigma is "a moral or physical blemish" ("Stigma," 1994, pg. 1004).

Assumptions

The current study includes several assumptions. One assumption was that the participants understood the variety of causes of HIV/AIDS. Another assumption was that conspiracy was suitable as a phenomenon for exploration in the study. A third assumption was that a stigma exists regarding the phenomenon.

The influence of historical behaviors by the medical and scientific community on African-Americans' sexual health decisions was an assumption of the study. The movie *Precious*, based on a true story, depicted this assumption

(Winfrey et al, 2009). When the main character's mother learns of her HIV status, she articulates that she thought she could not have contracted HIV because she did not have anal sex with her AIDS-infected husband.

Scope, Delimitations, and Limitations

SCOPE

The scope of the study consisted of interviewing African-Americans living in various states, including Hawaii. The basis for the selection was primarily the disproportionally prevalent incidence of HIV/AIDS among Americans of African descent when compared to members of other ethnic groups and residents of other countries.

DELIMITATIONS

In research, delimitations are decisions researchers make about the scope of the research. The current study did not include Africans from the continent of Africa because travel was a barrier and the focus remained on African-Americans. The study sample did not include African-Americans from every state, given time constraints. The data collected included only the participants' (a) perceptions and beliefs about conspiracy theories and sexual health decisions in the context of disproportionately high number of new cases of HIV among African-Americans and (b) perspectives on genocide associated with the Tuskegee experiments. The line of inquiry was limited to examining opinions related to conspiracy ideas, perceptions of genocide, and sexual health choices.

Some logistical concerns were delimiting factors. Most African-Americans familiar with the Tuskegee experiments reside in the southeast region of the United States and are typically older. No delimitations existed in the following areas: sexual orientation, sexual activity, sexual experiences, gender, or sexual identity (i.e., transsexual, unisexual, or intersexual).

LIMITATIONS

Research integrity is the foundation to support the research process. Some issues confronted in research are limitations outside the researcher's control. Limitations may or may not have any influence on the study, depending on the research goals. Few limitations strengthen the integrity of a study (Leedy & Ormrod, 2005). One limitation discovered while conducting this study was not being able to travel across the country to interview participants where HIV/AIDS is most prevalent. Budgetary constraints prohibited this approach to data collection. Another limitation discovered while conducting the study pertains to scheduling time with the participants for face-to-face interviews. For this reason, most of the interviews were conducted by telephone.

Being unaware of participants' HIV status limited the researcher's ability to garner a perspective from someone actually infected with HIV. Also, the participant selection process, less randomized, limited the researcher's ability to generalize across broader populations. Although one could argue that any research data reporting system is compromisable, the researcher's interview approach left questions relative to how truthful participants' responses were. Cultural bias may have influenced the researcher's interrelations with and interpretations of participant responses.

The qualitative phenomenological nature of the study was another limitation. Qualitative researchers largely explore questions about shared human occurrences (Ary et al, 2006). Qualitative research is generally designed to measure human experiences, which is subjective material, so the meaning of the study findings may not be applicable to other groups. The small sample in the study was not a limitation because data saturation occurred (Mason, 2010).

The study participants were selected because of their in-depth experiences regarding the phenomenon; the interviews allowed for extrapolating in-depth, comprehensive information about the phenomenon. Some limitations included restraints in time to find more potential participants. The location of the interviews coincided with the geographic region of many participants. Other individuals participated in interviews by telephone.

While only marginally captured while conducting the

study, fear was represented in the guise of stigma, shame, and ostracism. Such fear was not further analyzed to determine if participants were escaping seeking health information, a concept coined by the University of Florida researchers as "health information avoidance" (Hoffman, J., 2013). This limitation can be viewed as a gap in the body of existing research on HIV/AIDS.

As is appropriate in qualitative, phenomenological research, hermeneutic, interpretive dialogue produced the data. This approach eliminates limitations inherent in quantitative data.

The qualitative data collected in the study led to crucial findings. Leaders may use the narrative data to address the problem of misperceptions and the high prevalence of HIV/AIDS in the African-American community and to promote successful outcomes.

Summary

Chapter 1 contained a synopsis of the background of the problem examined in the current study. Topics included (a) the problem and purpose statements; (b) the significance of the study; (c) the nature of the study; (d) the research questions; (e) the theoretical framework; (f) the assumptions; and (g) the scope, delimitations, and limitations. The problem statement contained information on the general and specific problems addressed in the study, indicating the urgent need for the study. Both HIV and AIDS continue to spread quickly, especially among African- Americans.

Staff at the CDC (2007b) posited, "[T]he estimated number of HIV infections in adults and adolescents in the 50 states and the District of Columbia in 2006 was 50,000…22.8 per 100,000" (pg. 7). According to the CDC, the estimated number of AIDS cases was 37,041 as of 2007. Contracting HIV/AIDS is more likely when sexually transmitted infections (STIs) are present (Sutton et al, 2009).

The presence of sexually transmitted infection (STI) provides important clues for detecting HIV/AIDS in the human body by a factor of two to five. The STI rates are also disproportionately higher for black people. Such high rates require urgent attention through adequate screening and treatment incorporated in comprehensive plans to swiftly decrease the high epidemiology within the African-American community (Sutton et al, 2009). Sexually transmitted infections can expose the presence of HIV in the human body, compromising the transmission rate. Methods to reduce the prevalence of HIV/AIDS in the African-American community should be prioritized through comprehensive plans enabling easy, stigma-free access to prevention, screening, and treatment aimed at halting STIs in general.

To examine the problem of HIV/AIDS among African-Americans, the objective of this study was to explore the lived experiences and perceptions of African-Americans in relation to making choices regarding sexual acts. The study included a specific focus on the potential influence of perceptions about genocide conspiracy and trust in the American health-care system on decisions regarding sexual behavior. A better understanding of the factors that influence choices about sexual behavior may assist leaders in developing better tools to combat HIV/AIDS.

The aim of Chapter 2 is to discuss scholarly research pertaining to the topic of the current qualitative, phenomenological study. Topics include definitions and classifications of HIV and AIDS, historical perspectives on HIV/AIDS, symptoms of HIV, groups of people particularly susceptible to HIV/AIDS, and the prevalence of HIV/AIDS among African-Americans. Other issues discussed in the literature review include the Tuskegee experiments, African-Americans and the U.S. health-care system, and social cognitive theory.

2

REVIEW OF THE LITERATURE

THE PURPOSE OF this qualitative, hermeneutic, phenomenological research study was to explore the perceptions, understandings, and opinions of African-American men and women pertaining to conspiracy theories, sexual health decisions, and HIV/AIDS exposure to treatment. The study provided data to gauge and further examine trends and perceptions on trust issues associated with perceptions of genocide. Chapter 2 contains a review of the literature on the topics addressed in the current study.

Topics include the history of HIV/AIDS, definitions of HIV and AIDS, and symptoms of HIV. The literature review contains a discussion of the demographics of the African-American population, HIV/AIDS in the African-American population, African-American subpopulations with a high number of infected members, and potential causes of the prevalence of HIV/AIDS among African-Americans. Gaps in the literature are another essential feature of Chapter 2, supporting the need for more research in specific areas.

Keyword Searches and Documents

Keyword searches conducted in major databases, such as ProQuest, EBSCOhost, and MEDLINE, and in search engines like Google, generated literature pertinent to the current study. Historical and current literature provided a comprehensive view of the topics, including the prevalence of HIV/AIDS in the African-American population and the subgroups most afflicted, African-Americans' perceptions of medical professionals and experiences related to medical care, the Tuskegee experiments, perceptions regarding the genocide conspiracy theory, and the resulting distrust of the medical community. The literature discussed in the review includes peer-reviewed journal articles, newspaper articles, government reports, dissertations, and books.

History of HIV/AIDS

The first case of HIV/AIDS was detected in America in 1981. The incidents of HIV/AIDS for black Americans was 0%. Since then, HIV/AIDS has become a persistent global problem. Behaviors that increase a person's odds of contracting and transmitting HIV include (a) drinking breast milk of an infected woman; (b) receiving contaminated blood transfusions and being exposed to contaminated blood byproducts; (c) engaging in intimate sexual contact and high-risk sexual acts, such as anal, vaginal, and oral sex, without protection; (d) having unprotected sex with multiple partners; and (e) using and sharing contaminated hypodermic needles. Transmission of HIV does not occur through kissing, insect bites, casual contact, sharing food, drinking from the same glass, toilet seats, or coughing (CDC, 2008). The CDC (2010) reported that health-care disparities, racism, lack of education, poverty, and other social factors are barriers to care for the poor and minorities.

Limited data are available about the origins of HIV/AIDS. The disease afflicts every ethnicity, but African-Americans represent a disproportionate number of cases compared to the rest of the American population. Wolfe (2011) noted,

> The virus was in humans for 50 years before it spread widely. It spread another 25 years before French scientists Francoise Barre-Sinoussi and Luc Montagnier, who would go on to win a much-deserved Nobel Prize for their work, finally discovered it. (pg. 13)

The retrovirus of HIV/AIDS had infected human beings long before screening was developed. Wolfe (2011) discussed scientists' efforts to identify the etiology of HIV/AIDS over several decades to document how the virus started. Initially, HIV/AIDS was labeled incorrectly a gay white male disease. Cases have been identified in many people and have ravaged the African-American community for over a decade. According to Wolfe,

> Our conclusion was that these locations likely showed what HIV looked like prior to its global spread. We learned that people in these rural villages had an incredible and intimate level of contact with wild animals. The process of butchering involved direct contact with virtually all of the blood and body fluids that viruses call home. As we expected, people who were engaging in hunting and butchering were at the front line of viral transmission from animals to humans. Essentially, following the entry of the virus from chimpanzees in the early twentieth century, it likely maintained itself in small villages. (pg. 109)

The HIV/AIDS virus was identified initially in isolated small villages in Cameroon, Africa. The virus jumped from animals to humans as a result of widespread butchering of chimpanzees for bush meat and spread pandemically across the globe with the development of global travel.

The disease referred to as HIV/AIDS is most prevalent among the people of the continent of Africa, but people of all continents and countries are susceptible to HIV/AIDS. Between 1981 and 1993, HIV/AIDS among blacks was most prevalent in the male gay population, but the number of people diagnosed with the disease among all African-Americans began to increase. In 1993, a peak occurred, with approximately 70% of all HIV/AIDS cases identified among all U.S. ethnic minority groups, including Asian-Americans, American Indians, Hawaiians, and other Pacific Islander groups (CDC, 2007a).

A 2008 CDC report indicated stark differences in infection rates for different races. Caucasians in Minnesota compose 88% of the state's population but only 53% of new HIV/AIDS diagnoses in the state. African-Americans make up 3% of the overall population in Minnesota but 21% of newly diagnosed cases of HIV infection. These numbers contrast somewhat with national trends, in which African-Americans represent more than 50% of newly infected cases of HIV/AIDS (CDC, 2008).

According to the National Association of Social Workers (NASW, 2012), "Concerns about stigma affect an individual's decision to get tested, access health care, and withhold information about their status from family members, friends, and care providers" (pg. 4). Stigma and shame are of particular concern for minorities and the poor, populations that routinely experience discrimination. Concern about stigma influences decisions to seek screening, protection of self and others, and health care. More research on stigma among minorities and the poor could aid in decreasing the epidemiology of HIV/AIDS.

HIV/AIDS Path from Africa to America

During the course of conducting interviews for this research, many of the 16 participants asked questions related to the origin of HIV/AIDS, a disease that produces two million new infections every year worldwide. The origin of HIV/AIDS is analogous to the question "Which came first, the chicken or the egg?" This book attempts to provide information on both sides of the issue. Some scientists believe the origin lies within the scientific realm. Conspiracy theorists speculate otherwise.

Dr. Monica Gandhi (UCTV, May, 2013) provided a comprehensive historical account of the origin of HIV/AIDS. In the early years, AIDS was referred to in Africa as "slim disease." Dr. Gandhi posed the question "How do new infectious diseases enter the human population?" She summarized the answer this way: HIV/AIDS was once an "emerging disease" that entered the human species. Diseases emerge as a result of entering a new host (such as in humans, monkeys, chimpanzees) as a new infection. Dr. Gandhi refers to HIV as a "zoonodic" infection or "zoonosis," as in originating from animals, meaning the infection has emerged with the capability to jump species from animals to humans. Whenever humans interact with animals, they are exposed to a variety of contagious diseases like HIV/AIDS.

One important fact about slow-moving viruses (Lenti viruses) such as HIV and SIV, found in primates, is that there are many sub groups. Dr. Gandhi pointed out there are five major lineages of primates that host SIV or simian immunodeficiency virus. They are (a) chimpanzee (b) monkeys; mandrills (c) African green monkeys, baboons (d) sooty mangabeys, and (e) Syke's monkeys (Hahn, 2000). There are two major types of HIV in humans—HIV-1 and HIV-2—and many subtypes; however, the major four groups of HIV-1 are: group M, group N, group O, and group P. Group M causes 90% of all human infections. HIV-1, group M, strain B (clade) is known as the pandemic strain (Hahn, 2000). Most notable, Dr. Gandhi pointed out that HIV-1 is the SIV strain most frequently identified in the common chimpanzee (Pan troglodytes troglodytes). HIV-2 is most frequently found in sooty mangabeys.

How did the virus jump species, from primates to humans?

Americans were stricken with polio in the early 1950s, and scientists raced to design the most effective, safe vaccine for the disease. Polio vaccines, often concocted from monkey kidneys, were very effective in eradicating the disease. According to Gandhi (2013) there was competition in the 1950s among scientists Albert Sabin, designer of the oral vaccine for polio, and Jonas Salk, designer of the injectable vaccine for polio. Hilary Koprowski competed against Sabin for the first oral vaccine and lost. Despite ethics, Koprowski went on to administer his version of the oral vaccine in Belgium; however, he did not use chimpanzee cells. The crossover events include two prevailing theories: (1) Frequent contact between humans and primates in the "bush meat" trade for food and during the hunt for wild, exotic game. (2) Blood exposure to the hunters of "bush meat." These hunters' DNA contained a valley of SIV strains (Wolfe, N. 2011). The first case of HIV in the United States of America occurred in 1981. One case report from the Lancet (1959) describes a 25-year-old male Navy seaman with several symptoms mimicking HIV such as emaciation, ulcerative upper lip, and pneumocystis in his lungs. Postmortem results concluded he had HIV/AIDS.

Most cases of HIV/AIDS originated in West Africa. Previous to 1981, there were no regional cases of HIV reported. The oldest known specimen stored in history was in Africa in 1959. Among 1,213 stored specimens, only one of those was HIV-1. Examining stored human specimens from previous centuries is the only true means of deciphering how far back in time HIV goes.

HIV/AIDS Definitions and Criteria

Schneider et al (2008) provided a definition of HIV/AIDS and the criteria for its identification. The specific HIV/AIDS classification system included adults, adolescents, and children under 18. Schneider et al's report included notes pertaining to case definition that must include laboratory results positive for the following conditions:

> Bacterial infections, multiple or recurrent; Candidacies of bronchi, trachea, esophagus, or lungs; Cervical cancer, invasive; Coccidioidomycosis, disseminated or extra pulmonary; Cytomegalovirus retinitis (with loss of vision); Encephalopathy, HIV-related; Herpes simplex: chronic ulcers (> 1 month's duration) or bronchitis, pneumonitis, or esophagitis (onset at age > 1 month); Histoplasmosis; Kaposi's sarcoma; Lymphoid interstitial pneumonia or pulmonary lymphoid hyperplasia complex; Lymphoma, Burkitt's (or equivalent term); Lymphoma, immunoblastic (or equivalent term); Lymphoma, primary of brain; Mycobacterium avium complex or Mycobacterium kansasii, disseminated or extrapulmonary; Mycobacterium tuberculosis of any site; Pneumocystis jirovecii pneumonia; and Wasting syndrome attributed to HIV. (para. 6)

Revised definitions included CD-4+ T lymphocyte counts and percentages added to the objective laboratory conclusions (CDC, 2008). The disease cannot be diagnosed solely on the basis of phenotypic observation, and qualified medical personnel only must conduct laboratory testing.

HIV TREATMENT

Medical treatment is known to decrease symptoms associated with HIV. Clinical trials in Thailand indicated that the HIV vaccination is effective. McGrath (2009) reported that 16,000 people in Thailand received a combination of two earlier experimental vaccines. Individuals receiving the vaccine were 31.2% less likely to contract HIV.

Current research shows that people with HIV are living longer because of new antiretroviral medications, but exhibit serious psychological symptoms. People living with HIV tend to rely on taking pills and visiting doctors rather than making wise sexual health choices (McGrath, 2009). Little to no research exists on whether interventions to address these issues have helped decrease the number of new cases of HIV/AIDS in the African-American population.

The members of the 2012 Consultation on Revision of the HIV Surveillance Definition convened from 2008 to 2011 (CDC, 2012). Several workshops culminated in the recommendation to revise certain aspects of the surveillance case definition to include (a) new comprehensive HIV testing algorithms, (b) a definition of HIV-2 infection that differs from the familiar HIV-1 definition, (c) a change in staging nomenclature to include HIV-0 with specific virologic criteria or serologic criteria defining antibody test results, (d) inclusion of opportunistic illnesses in the third stage and inclusion of T-cell results when determining staging for both adults and adolescents, and (e) staging criteria with elimination of the term AIDS to reference stage-3 HIV infection in surveillance reporting.

Other changes included revising the HIV infection surveillance case definition for children, and adding criteria for specific physician-documented diagnosis classifications. For example, when physicians converge on a diagnosis for HIV, they should enter the diagnosis as "laboratory test-documented, rather than physician-documented" (CDC, 2012, pg. 2). The members of the 2012 Consultation on Revision of the HIV Surveillance identified the following steps as (a) submission of recommendations in a position statement to the June 2012 meeting of the Council of State and Territorial Epidemiologists (CSTE); (b) revision of the proposed case definition based on further recommendations from the CSTE; (c) publication of the accepted revisions in an MMWR in 2012; and (d) implementation of the revisions by changing case report forms, database software, and reporting laws and practices by 2013.

The newer, stricter, and better-defined criteria for HIV may impact surveillance trends. A more scientifically defined nomenclature might improve clinical trials. The new definitions should reflect more objective measures for classifying HIV and AIDS in adults and children. Such changes might enhance research efforts to tailor clinical trials, differentiate between stages, and create more effective medications for HIV (CDC, 2012).

The Affordable Care Act addresses the issue of lack of care for the poor and disadvantaged populations. The act includes individuals with preexisting conditions like HIV/AIDS. The legislation paves the way for the basic privilege of health insurance to cover (a) preventive services; (b) treatment for conditions of longstanding origins; (c) coverage for children through age 26, many of whom may be reluctant to screen for HIV/AIDS or are turned down for treatment for lack of insurance; and (d) a wider range of coverage for individuals and families, including those with incomes significantly lower than the poverty level (American Public Health Association, 2012).

HIV/AIDS AND STIGMA

Initially, HIV/AIDS was found mostly among gay white men, and few African-Americans were reporting symptoms (CDC, 2006). The epidemiologic map has changed. For over a decade, the epidemiology of HIV/AIDS among blacks has remained consistently higher than among all other ethnic populations in the United States.

Stigma and shame are barriers to screening, education, prevention, and treatment. Tomaszweski wrote an article in the NASW (2012) describing how stigma and shame of HIV/AIDS among minorities and the poor create social obstacles to health surveillance and prevention tactics. Tomaszweski noted that perceptions of stigma and shame influence minorities to believe they are not at risk for HIV/AIDS.

SYMPTOMS OF PROGRESSIVE HIV

Many African-Americans lived for months, sometimes years, with HIV/AIDS before obtaining a diagnosis (Blankenship, Smoyer, Bray, & Mattocks, 2005). Stigma, shame, and perceptions of conspiracy may often serve as barriers to conscious, healthy sexual choices. Preventive measures are ideal to avoid contracting HIV/AIDS, but awareness of and alertness in identifying symptoms are important behaviors, so that swift action can be taken if symptoms of the virus are present.

The most objective measures for identifying the presence of HIV/AIDS are FDA- approved tests that measure the presence of specific antibodies. Conventional HIV/AIDS tests include the HIV ELISA/Western Blot, a set of blood tests used to detect the presence of antibodies in the blood. A person's unique immune response determines his or her strength when the symptoms manifest. In some cases, HIV can lie dormant in the bloodstream for many years.

For many individuals, regardless of ethnicity, becoming infected with HIV presents without symptomology, persisting this way for as much as a decade before any illness state manifests. This situation provides a window of opportunity to transmit the virus easily and unknowingly. After exposure, detection of HIV through screening takes approximately three months (CDC, 2009).

Health-care personnel who provide testing and screening measures might best serve patients suspected of harboring HIV/AIDS by participating in continuous education so they have updated information on symptomology. Up-to-date knowledge helps professionals accurately educate patients and recognize when a formal assessment is warranted. A CDC (2006) report indicated, "Health-care providers should be knowledgeable about the symptoms and signs of acute retroviral syndrome, which is characterized by fever, malaise, lymphadenopathy, and skin rash" (pg. 2). Everyone, including African-Americans, needs to (a) understand the symptoms and disease states of HIV/AIDS, (b) be open and honest about sex for effective communication and adequate protection against the exchange of bodily fluids, and (c) be more vigilant about looking for symptoms that might indicate HIV/AIDS.

EPIDEMIOLOGICAL RATES OF HIV/AIDS

According to the Department of Health and Human Services Centers for Disease Control and Prevention (CDC, 2012), African-Americans face the most disparity and burden among U.S. racial groups. Despite representing only 13% of the U.S. population in 2009,

African-Americans accounted for 49% of all new HIV infections in that year. Numerous disturbing facts in the same report supported a need for the present phenomenological research study. The facts include (a) young African-American gay and bisexual men are especially at risk of HIV infection; (b) in 2009, black men accounted for 70% of the estimated new infections among all blacks; (c) the estimated rate of new HIV infection for black men was more than six and a half times as high as that of white men, and two and a half times as high as Latino men and black women; (d) in 2009, black men who had sex with men (MSM) represented an estimated 73% of new infections among all black men, and 37% were MSM, leaving 36% represented by straight black men; (e) in 2009, black women accounted for 30% of the estimated new HIV infections among all blacks; (f) most (85%) black women acquired HIV through heterosexual sex; and (g) the estimated rate of new HIV infections for black women was more than 15 times as high as the rate for white women, and more than three times that of Latino women (CDC, 2009).

HIV/AIDS in the African-American Population

The 2010 census provided new data for African-Americans. The estimated number of African-Americans was 40 million, comprising "12.3 percent of the U.S. population, down from percent of the population in

2000" (Bowman, 2010, para. 1). Out of 37,131,771 African-Americans, 47.6% are men and 52.4% are women (Bowman, 2010). In America, one in five African-American families lives in poverty. Poverty affects access to health care and may contribute to the prevalence of HIV/AIDS in the African-American population. Since HIV/AIDS is a preexisting condition, the epidemiology of HIV/AIDS may be a negative factor for all Americans lacking health care.

DISPROPORTIONAL RATES IN THE AFRICAN-AMERICAN POPULATION

Research indicates that various populations have a higher prevalence of HIV/AIDS than other populations (CDC, 2008). African-Americans have a higher incidence of new cases of HIV/AIDS; the disparity has persisted for about a decade. The disproportionately high rate of HIV/AIDS in the African-American community has received great attention since the beginning of the 21st century.

African-Americans of all ages account for 51% of the 42,655 new HIV/AIDS diagnoses in 34 states with long-term reporting (see Figure 1). African-Americans account for 48% (with long-term reporting) of the 551,932 individuals of all ages living with HIV/AIDS in 34 states. The rate of AIDS diagnosis for African-Americans is ten times the rate for Caucasians and nearly three times the rate for Hispanics.

GENDER AND HIV/AIDS

As shown in Figure 2, the rate of HIV/AIDS diagnosis in African-American women is 22 times the rate for white American women. The rate of AIDS in African-American men is almost eight times the rate for Caucasian men. By the end of 2007, 40% of the 562,793 persons with HIV/AIDS who died were black (CDC, 2009). For African-American women with HIV/AIDS, the most common means of transmission was high-risk heterosexual contact (e.g., sexual contact with men with HIV/AIDS and use of drug injection needles shared with someone infected with HIV/AIDS) (CDC, 2009).

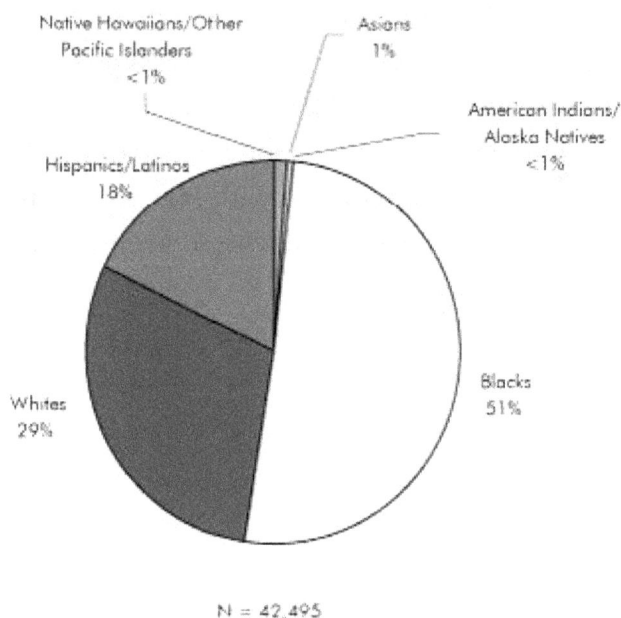

Figure 1. Incidence of HIV in various racial groups from 33 U.S. states (CDC, 2009).

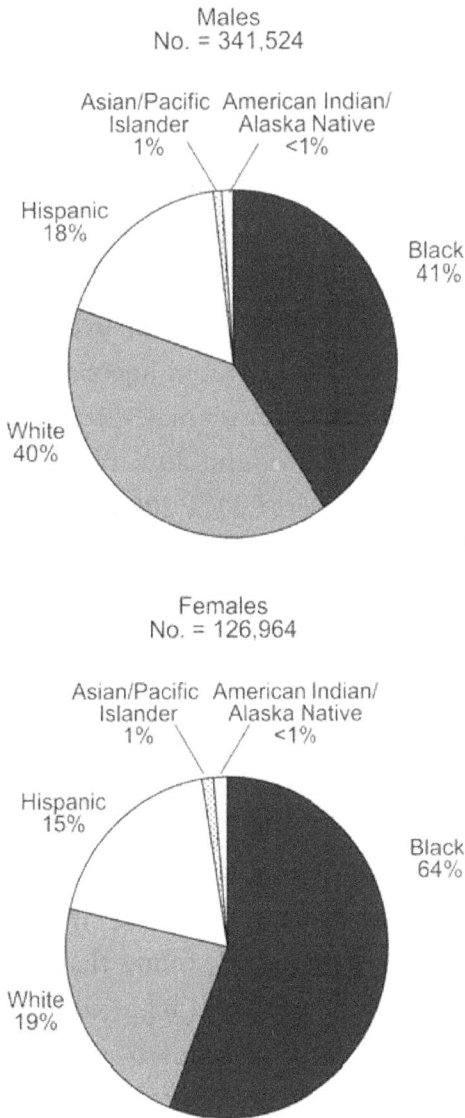

Figure 2. Incidence of HIV by racial group and gender in 33 U.S. states (CDC, 2009).

Statistics indicate that higher percentages of African-American men are in high-risk groups diagnosed with HIV. "Black males accounted for more new HIV/AIDS diagnoses than males of any other racial/ethnic population in the South (47.5%)" (CDC, 2007a, pg. 190). Between 2003 and 2007, the incidence of new HIV/AIDS diagnoses in the African-American male population was categorized as follows: 50% resulted from male-to-male contact, about 20% resulted from injection drug use, approximately 10% resulted from male-to-male contact and injection drug use, and around 20% resulted from high-risk heterosexual contact (CDC, 2005). A CDC (2007a) report indicated that 64% of African-American women in high-risk groups tested positive for HIV. Two-thirds of newly reported cases of HIV were in African-American women (CDC, 2007a).

African-American men represent a growing sub-population of African-Americans contracting HIV/AIDS. Bing, Bingham, and Millett (2008) noted, "Male African-Americans account for 71% of all reported HIV/AIDS cases among African-Americans" (pg. 244). Siegel, Schrimshaw, and Karus (2004) found that some African-Americans continue engaging in high- risk sexual activity after receiving news of being infected with HIV.

A crucial gap in the literature concerns heterosexual African-American men's views on and knowledge about HIV/AIDS (Thompson-Robinson, Weaver, Shegog, & Richter, 2007). Examining African-American men's perceptions and understanding of HIV/AIDS will help researchers, leaders, and medical professionals understand the reasons for the sexual health choices of the majority of African-American men. Focused research and accurate assessments can lead to important insights about the reasons why some African-American men choose not to practice safe sex.

Men who have sex with men (MSM) are considered to be a part of the sexual and gender identities population, according to the CDC (2007a). Bing et al (2008) reported that African- American men account for more than half of unprotected sexual activities with other men. Among African-American men, black homosexuals continue to have the highest prevalence of newly diagnosed cases of HIV/AIDS (CDC, 2007a). This condition indicates that reaching out to and educating homosexual African-American men about HIV/AIDS and preventing and treating HIV/AIDS in this subpopulation remain difficult challenges.

Despite the prevalence of homosexuality among African-American men, homosexual activity, especially between men, remains taboo and stigmatizing in the African-American community. Men who engage in unprotected sex with other men are at extreme risk of contracting HIV (Peterson & Jones, 2009). The stigma associated with homosexuality in the African-American community may result in a vicious cycle of fear and denial, leading African-American homosexuals to make poor sexual health choices.

Increased epidemiology of men of color who have sex with other men (MSM) of color reflects targeted stigmatization from leadership, stemming from federal policy orchestrated by Jesse Helms in 1987 encouraging the omission of MSM from funded programs ("MSM of Color," 2011). Lawmakers must propose legislation inclusive of all people when forging policy for education, screening, and treatment of HIV. Given the higher rates of infection within the African-American and MSM communities, leaders and policy makers should encourage recruitment of more minorities in health care and more funding efforts to fight the battle of HIV/AIDS.

PRISON POPULATION

A range of research on African-Americans in the prison system provides disturbing statistics on the disproportionate number of inmates with HIV/AIDS. Proposed causes of the prevalence of HIV/AIDS among inmates have reflected social context and structural perspectives. Factors leading to the prevalence of HIV/AIDS in this population include poverty, homelessness, community disintegration, lack of access to services for sexually transmitted diseases, discrimination, and racism (Blankenship et al, 2005). These social factors compound the dilemma of HIV/AIDS in the prison population with the absence of progressive opportunities for equal opportunities for jobs, housing, and quality education and training as deterrents to crime.

The prison population reflects a higher percentage of incarcerated minorities compared to other ethnicities. African-Americans are disproportionately represented in the prison population; 10% of inmates in the prison system are African-American (Cornelius & Hamilton-Mason, 2009). Data are lacking on the prevalence of HIV/AIDS among African-American prisoners compared to prisoners of other races. Risky behaviors are reported to be more frequent among African-Americans than in other groups, indicating a possible higher incidence of HIV/AIDS among African-American inmates (Blankenship et al, 2005). Risky sexual practices in prison contribute to a higher incidence of HIV/AIDS. High-risk sexual practices, such as men having sex with other men in the prison environment, create a serious problem for disease control in the general population. When inmates exit the prison system, they may engage in sexual activity with new individuals and transmit HIV/AIDS and other diseases.

SUBSTANCE USERS

Drug users who share intravenous needles are at an increased risk of contracting HIV/AIDS. Such practices have prompted needle exchange programs in Hawaii and South Carolina. Research indicates such programs are effective in reducing HIV/AIDS among drug users, but leaders of many major cities continue to ban needle exchange programs. Opponents argue that the programs only lead to increased use of clean needles, which may contribute to higher use of illegal drugs. If drug users cannot afford to use clean needles, they use needles that may have been contaminated, because their desire for illegal substances is greater than their concern for their health.

Another consideration of substance use is that it decreases inhibitions about sex and is driven by poverty, mental illness, a lack of opportunities, and unemployment (Peterson & Jones, 2009). Addressing these underlying factors may reduce the number of drug users. A decrease in the number of substance users may lead to a decrease in the number of Americans who contract HIV/AIDS.

Causes of the Prevalence of HIV/AIDS in the African-American Population

Research indicates many factors may contribute to the prevalence of HIV/AIDS among African-Americans, including historical factors that have caused African-Americans to believe in conspiracy theories and distrust the U.S. medical community (Bogart & Bird, 2006). Wyatt et al (2009) reported other social forces include homelessness, incarceration rates, unemployment, the lack of communication with sexual partners, misinformation about the spreading of HIV, and lack of access to health care. The following subsections include a discussion of various factors.

HISTORICAL RELATED FACTORS

The history of African-Americans in the United States is long and tumultuous, spanning over 300 years. Throughout much of this history, the relationship between some African- Americans and Caucasians has been filled with tension and contention. Much of the mistrust may have emanated from a long history of discriminatory practices, particularly two key historic events: slavery and the Tuskegee experiments.

These two events may have led to conspiracy theories among African-Americans that HIV/AIDS was developed to hurt the African-American population. Such perceptions are more common among African-Americans born before 1990 (Bogart & Bird, 2003). Dispelling theories of conspiracy and genocide may be an incredibly difficult task, but may reduce the barriers to unhealthy sexual behavior. Bogart and Bird (2003) noted, "Almost no research has investigated the relationship between such beliefs to behaviors and attitudes relevant to HIV risk" (pg. 1057).

Slavery. Campinha-Bacote (2009) noted that "African-Americans are largely descendants of Africans who were brought forcibly to the United States as slaves between 1619 and 1860" (pg. 49). Slavery ended with the ratification of the 15th Amendment to the Constitution on February 3, 1870. African-Americans are now the largest ethnic minority group in America. Since slavery ended, African-Americans have advanced in many areas, and legislation has been passed to support fairness and equality in response to ongoing struggles for equal rights. Americans can now boast of having the first African-American president, Barack Obama. Nevertheless, slavery from the 16th to 19th centuries left a blemish on the country and has remained a cause of distrust between African-Americans and Caucasians (Campinha-Bacote, 2009).

The Tuskegee experiments. The Tuskegee experiments, combined with the long history of slavery in the United States, may have led to conspiratorial theories among African-Americans, such as HIV/AIDS being perceived to have been developed to obliterate the African-American population and other minority populations. Such perceived beliefs may have subconscious bearings on the sexual health choices of African-Americans. Exploring perceptions that center on past injustices and questionable ethical practices may be warranted. According to Bogart and Bird (2003), "News accounts and a small but growing amount of public health literature have reported widespread belief in conspiracy theories about the U.S. government and health care systems among Blacks, such as beliefs related to HIV, birth control, and genocide" (pg. 5). Some degree of support might exist that corroborates theories regarding a history of unethical scientific health practices among blacks in America.

The Tuskegee experiments were part of a study conducted by American physicians and scientists between 1932 and 1972. The study was conducted to determine the effects of untreated syphilis on African-American men, and became the most egregious assault on African-Americans since the atrocities of slavery (Lombardo & Dorr, 2006). Racism, bureaucratic inertia, and the personal motivations of study personnel appeared to have led to the absence of ethical practices associated with the Tuskegee experiments that allowed such studies to persist for four decades.

Surgeon General Hugh Cumming and two assistant surgeon generals, Taliaferro Clark and Raymond A. Vonderlehr, the study's principal investigators, were graduates of the University of Virginia, where training on the treatment of syphilis was based on race and genes. The University of Virginia was noted for medical studies in the field of eugenics and, according to Lombardo and Dorr (2006), the Tuskegee experiments "provided the vehicle for testing a eugenic hypothesis: that racial groups were differently susceptible to infectious diseases" (pg. 291). The lack of ethics resulted in the Tuskegee experiments expanding into a four-decade odyssey in search of the truth on genes and disease. The disregard for the human participants in the Tuskegee experiments led to African-Americans' distrust of the U.S. health-care system.

Mistrust of the U.S. health-care system. The slavery experience and the Tuskegee experiments may be factors from historical events that contribute to why African-American men visit doctors far less often than Caucasian men (King, 2003). Not having access to health screenings, or refusing screenings, presents serious barriers for containing the spread of the HIV/AIDS virus. W. D. King (2003) noted,

> Recent analyses of the relationships of minority patients with their physicians have demonstrated that provider racism and patient awareness of invidious past events such as the experimentation on slaves and the Tuskegee syphilis experiment have contributed to minority patients having less access to and knowledge of specific medical treatments than their white counterparts, 5–10 lower levels of trust, and greater unwillingness to participate in clinical trials. This places minority patients at a disadvantage not only in receiving preventive care but also in access to newer technology and treatment stemming from clinical trials. (pg. 366)

W. D. King's (2003) assertion might indicate that African-Americans' distrust of the U.S. health-care system may be widespread, and a significant factor when considering the battle against HIV/AIDS and the disproportionate number of new cases among African-Americans. Distrust may also be a barrier to African-Americans deciding to be screened for HIV/AIDS, even though not knowing one's

HIV/AIDS status before engaging in any sexual act is extremely risky.

SEXUAL HEALTH CHOICES

Sexual health decisions may be as much a reason for the consistently high percentage of HIV/AIDS among African-Americans as is fear of stigma, conspiracy ideas, and the lack of health care. Wise sexual health practices include self-efficacy for regulating motivations and urges that drive sexual desires, using protection during sexual intercourse, abstaining from sex if protection is not available or comfortable, limiting the number of sexual partners, and refraining from engaging in high-risk sexual acts (Bandura, 1977). Other important sexual health practices include screening for HIV/AIDS, and insisting on knowing a sexual partner's HIV/AIDS status.

Multiple partners. Another factor contributing to the higher incidence of HIV/AIDS in the African-American community is the practice of having multiple sexual partners, who may be of the opposite gender or the same gender. Having multiple sexual partners is associated with increased HIV/AIDS risk as well as excessive use of substances, higher rates of incarceration, and mental distress (Mulatu, Leonard, Godette, & Fulmore, 2008).

Female-specific factors. One theory on the prevalence of HIV/AIDS in African- American women is that women are often subjected to abusive relationships, which may have contributed to a conscious lack of safe sexual health choices. Women who depend on men for support are a highly vulnerable group. According to Simoni and Ng (2002), African-American women, age 25 to 61, living with HIV/AIDS in New York City revealed high levels of sexual (39%) and physical (44%) trauma before age 16. Such trauma histories were implicated in perceived health status.

Social Cognitive Theory

Psychological perspectives regarding sexual health choices may be rooted in the principles of social cognitive theory. Social cognitive theory focuses on the psychological and social constraints and pressures that lead to specific choices and behaviors (Lazarous, Newman, & Shell, 2007). The application of social cognitive theory includes events leading to unsafe sexual practices (e.g., drug use and excessive drinking) that contribute to lessening the defenses of self-regulation.

Based on self-regulation and social cognitive theory, an individual's choice not to protect their sexual health may stem from a lack of conscious inhibitions or self-regulation. McCormick and Martinko (2004) found the process of self-regulation involves individuals taking charge to control and direct their actions. Human beings can become more goal-directed with their actions to engage safely in sexual encounters. Individuals could actively participate in certain practical patterns of thinking and behaving in response to goal-oriented ecological settings. Self-regulation is an important concept for considering education and training for HIV/AIDS and other sexually transmitted infections (STI) (McCormick & Martinko, 2004).

The concept of self-regulation is similar to Bandura's (1977) self-efficacy theory. With self-efficacy, individuals acquire the capacity to estimate as accurately as possible their ability to manage stress from day-to-day problems (Beck, Freeman, & Davis, 2004). Self-efficacy theory supports self-regulation, and is an essential element of the overall leader paradigm for ontological discourse on sexual activity and HIV/AIDS, and self-protection from STIs through self-regulation, from the perspective of a client's own reality.

When teens and women, in particular, become confident and acquire self-efficacy, they are better able to recognize the deception of intimate partners in the early stages of courtship. Women with self-efficacy feel far less inhibited about communicating their needs to protect themselves from sexually transmitted diseases, and can withstand pressure to engage in risky sexual behavior. Substance abuse may decrease individuals' self-regulation and self-efficacy.

Lazarous et al (2007) noted, "There is a body of research showing a consistent relationship between drinking behavior and condom use among young people. Most researchers agree condom use is affected by many factors" (pg. 44). Lazarous et al proposed that condom use, an action that occurs less often in individuals under the influence of substances, is uncomfortable to discuss during intimate moments and that abusing substances can make the act of following through on sexual desires more automatic. Despite this finding, additional research regarding the relationship between substance use and sexual inhibition and regulation is lacking.

Boer and Mashamba (2005) conducted a study to examine social cognitive theory in relation to sexual health choices. The researchers' goal was to determine whether social cognitive theory models support a cognitive basis for intended condom use among African-American adolescents. The study did not include an examination of the cognitive factor of mistrust in the American health-care system; further examination of this factor is necessary to better understand the full spectrum of issues causing the disproportionate epidemiological statistics of African-Americans with HIV/AIDS, especially with newly diagnosed cases. The epidemiology of HIV/AIDS remains higher for blacks in America (CDC, 2009).

Issues Regarding Research on HIV/AIDS

The high prevalence of HIV/AIDS indicates that more research is needed, particularly on the means of reducing the number of new cases and minimizing the spread of

HIV/AIDS for African-Americans. Since a dispropor-tionate number of African-Americans are diagnosed with HIV/AIDS, research on the African-American popula-tion is of particular importance. Wyatt et al (2009) noted, "Early career trajectories can limit the pool of African-American HIV/AIDS investigators, because students of-ten choose clinical careers before they fully understand research training options" (pg. 99). Research may also be lacking because young and eager scientists often have lim-ited knowledge of how to obtain research grants and find appropriate funding; these individuals may consequently focus on forensics and other clinically oriented arenas of science and medicine.

Individuals must take full responsibility for avoiding risky sexual behavior and maintain sexual health to pro-tect themselves and others against sexually transmitted diseases (STD), thereby decreasing the risk of contract-ing HIV/AIDS. Complacency, denial, and distrust of the medical community could continue to place African-Americans in danger of contracting HIV/AIDS. Other potential factors include a lack of health insurance and access to health care, poverty because of traditions of job discrimination, and incarceration. These variables may help explain the prevalence of HIV/AIDS among African-Americans (Blankenship et al, 2005).

The purpose of this hermeneutic, phenomenological research study was to explore the perspectives of African-American men and women on genocide conspiracy and sexual health practices in hopes of better understanding issues related to high rates of HIV/AIDS among blacks. The continuing prevalence of HIV/AIDS in the African-American population could indicate that research efforts have had a minimal impact on reducing HIV/AIDS among African-Americans. More research is needed, because HIV/AIDS is considered the eighth deadliest virus and a serious global health crisis with far-reaching humanitarian and medical implications, especially for people of African ancestry. The insight gained from the study may be used to develop interventions aimed at changing the perceptions of African-Americans and encouraging appropriate sexual health decisions, such as screening for HIV status. Such changes could lead to a decrease in the number of African-Americans who contract HIV/AIDS.

Health Administration Leadership

The concerns of health policy leaders and administrators about the epidemiological disparity of HIV/AIDS in the African-American community may need to be more sensitive to the history of relations between black people and the medical community. Medical leaders must focus on socioeconomic factors and health care deficits for poor and rural Americans, and provide education about where and how to get screening and treatment in geographic areas with a specifically high incidence of HIV/AIDS. Leaders combating HIV/AIDS must develop health administration policy focused on attitudes toward and treatment of patients with infectious disease, thus creating a new generation of health care professionals with knowledge about the spread and prevention of HIV, including an individual's HIV status.

Summary

Chapter 2 contained a review of the literature on several important issues pertaining to the current study. Topics included the definitions of HIV and AIDS, symptoms of HIV, historical perspectives of HIV/AIDS, and HIV/AIDS in the African-American population. Also included was a discussion of the history of African-Americans, the Tuskegee experiments, and African-Americans' consequent mistrust of the U.S. medical community. Sections of the chapter addressed African-Americans' use of the U.S. health-care system, social cognitive theory, gaps in the literature, and administrative and policy leadership. Chapter 3 includes an in-depth discussion of the research methodology utilized in the present study.

3

RESEARCH METHODOLOGY

THE PURPOSE OF the current qualitative, phenomenological study was to explore the perceptions and lived
experiences of African-Americans regarding the Tuskegee
experiments, the conspiracy of genocide, and sexual health
in relation to HIV/AIDS. The participants received a unique
opportunity to communicate their views on the phenomenon under study. The data obtained helped answer the research questions developed for the study. The objective for
the analysis of the data was to obtain a heightened awareness of African-Americans' experiences and perceptions of
factors that contribute to their sexual health decisions and
the prevalence of HIV/AIDS among African-Americans.

Chapter 3 contains a detailed description of the research methodology used in the current study. The chapter provides details of the research method and design,
and their appropriateness to accomplish the study goals.
Other topics include the study population and sample,
the sampling method, considerations regarding informed
consent and confidentiality, the data collection method,

measures used to ensure reliability and validity, and the data analysis process.

Research Method and Design

The qualitative approach and phenomenological design were selected as the best approaches to answer the research questions developed for the study. One section includes the rationale for selecting the qualitative approach. Another section addresses the nature and use of the phenomenological design.

QUALITATIVE METHOD

Creswell (2008) stated, "Qualitative research tends to address research problems requiring: An exploration in which little is known about the problem and a detailed understanding of the central phenomenon" (pg. 51). Qualitative research involves collecting subjective data consisting of perspectives and philosophical views. According to Creswell, the qualitative process involves a series of investigations based on inductive and deductive lines of reasoning to answer research questions. Researchers start examining a problem in broad scope and narrow the focus through a series of logical lines of inquiry (Creswell, 2008).

QUANTITATIVE METHOD

The quantitative method involves collecting numerical data that are subsequently analyzed with statistical tests. According to Creswell (2008), "Quantitative research is an inquiry approach useful for describing trends and explaining the relationship among variables found in the literature" (pg. 645). Many quantitative studies present statistical measurements of variables, such as the prevalence of a phenomenon, populations, age groups, ethnicities, geographic characteristics, and socioeconomic factors (Creswell, 2008). Quantitative research is valuable to test specific hypotheses. The quantitative approach would not have been appropriate to achieve the goals of the current study. Some scholars consider qualitative data less reliable than quantitative data. Other scholars claim that the information participants provide in qualitative interviews is often as valuable as objective, quantitative data. Qualitative research provides an opportunity for open dialogue and communication that leads to deep levels of understanding and solutions to social problems (Creswell, 2008).

The goal of the current study was to better understand the phenomenon of HIV/AIDS among African-Americans, particularly in relation to their experiences and perceptions of the Tuskegee experiments, the genocide conspiracy, and sexual health choices. The qualitative approach was appropriate for such inquiry. Scientists and leaders can use the in-depth, textual data the study participants offered to combat HIV/AIDS in the African-American population.

PHENOMENOLOGICAL DESIGN

Of the various qualitative research designs available, phenomenology was the most appropriate to obtain personally meaningful, insightful, and salient data about African-Americans' experiences and perceptions of sexual health choices, the Tuskegee experiments, and the genocide conspiracy. The phenomenological design supports the capture of multiple realities and varying perspectives and experiences regarding a phenomenon (Ary et al, 2006). Ary et al (2006) noted, "An experience has different meanings for each person" (pg. 31). The phenomenological design facilitates understanding the meanings of each person's unique experience. In the current study, the African-American participants shared their unique perceptions of the Tuskegee experiments and the genocide conspiracy that might have influenced African-Americans' sexual health decisions and the likelihood of contracting HIV/AIDS.

Phenomenological research often involves conducting interviews with open-ended questions so the participants have the opportunity to provide extensive responses about the topic of inquiry (Ary et al, 2006). Ary et al (2006) reported, "Interviews are used to gather data from people about opinions, beliefs, and feelings about situations in their own words. They are used to help understand rather than to test hypotheses" (pg. 438). Ary et al further noted,

> In a personal interview, the interviewer reads the questions to the respondent in a face-to-face setting and records the answers. One of the most important aspects of the interview is its flexibility. . . . The interviewer can also press for more information when the response seems incomplete or not entirely relevant. (pg. 380)

In this study, a carefully designed set of unstructured interview questions encouraged the participants to share their individual realities. All participants were African-Americans, a population with a high prevalence of HIV/AIDS. A diagnosis of HIV/AIDS was not a requirement for participation, but HIV/AIDS may have affected the participants, their families, and their acquaintances. The opinions of the study participants carried high value in understanding the central phenomenon.

Hermeneutics is frequently associated with theological interpretation of the Bible. Erdi, Ujfalussy, and Diwadkar (2009) explained the relationship between phenomenology and hermeneutics and stated,

> Hermeneutics is a branch of continental philosophy which treats the understanding and interpretation of texts. Philosophical hermeneutics emphasizes existential understanding instead of

interpretation. Critical hermeneutics offers a methodologically self-reflective reconstruction of the social foundations of discourse and inter-subjective understanding. Finally, phenomenological hermeneutics is an attempt to synthesize the various hermeneutic currents. (pg. 413)

In hermeneutic phenomenology, the researcher explores participants' ontological perspectives while using interpretive reconstruction of the discourse to understand the lived experiences in the present (Erdi et al, 2009). Hermeneutic phenomenology was appropriate for the study to fulfill the goals to explore, interpret, and understand the sexual health decisions of African-Americans and the potential influence on such decisions of perceptions regarding the Tuskegee experiments and the genocide conspiracy. The analysis and interpretation of extensive data collected through the one-on-one interviews generated understanding of the varying perspectives and experiences about the central phenomenon. The analysis and interpretation of the data resulted in answering the research questions:

R1.   What are the lived experiences and perceptions of African-Americans pertaining to conspiracy beliefs, and the influence of such beliefs on their sexual behaviors?

R2.   What influence, if any, do conspiracy beliefs held by African-Americans have on their decisions associated with seeking medical attention?

Study Population, Sample, and Sampling Method

Ary et al (2006) defined the term *population* as including "those about whom you wish to learn something" (pg. 54). The population chosen for the current study consisted of African- American men and women who were from 18 to 65 years old and lived in Florida and Hawaii. The individuals may have been directly or indirectly affected by HIV/AIDS. The selected population was appropriate because the goal of the study was to explore African-Americans' perceptions and experiences regarding the Tuskegee experiments, the genocide conspiracy, and sexual health behaviors in relation to HIV/AIDS. Participation was not restricted to individuals infected or not infected with HIV/AIDS because the intent of the study was to understand the experiences of various members of the African-American population rather than limiting exploration to individuals who did or did not have HIV/AIDS.

The recruiting of participants took place by mail and telephone contact. Potential recruits received referral forms through the United States Postal Service (USPS) or at a health fair during the summer. Some participants were referred by word of mouth from other participants engaging in the research study. Many participants were service personnel from various states who were stationed in Hawaii. Other participants were individuals who had been transplanted to Hawaii from the mainland. Participants residing out of state communicated by mail and telephone.

Informed Consent

Creswell (2008) explained informed consent as a method of respecting an individual's commitment to the research process by explaining the individual's rights associated with participation in the study. Before initiating data collection, each participant received an informed consent form (see Appendix B). The participants interviewed by telephone received the form by e-mail before the interview; participants who completed in-person interviews completed the form at the time of the interview.

The informed consent form contained an explanation of the study purpose, the data collection process, the voluntary nature of participation, and the participant's right to withdraw from the study at any time. The informed consent form contained a description of the potential risks and benefits of participation, and notification that the interviews would be audio-recorded with the participant's permission. The participants received assurances that their participation would remain confidential (Foster & Cone, 1998). The informed consent form included the following statement:

> By signing this form you acknowledge that you understand the nature of the study, the potential risks to you as a participant, and the means by which your identity will be kept confidential. Your signature on this form also indicates that you are 18 years old or older and that you give your permission to voluntarily serve as a participant in the study described.

Each participant signed the form before participating in an interview.

Confidentiality

The application of guidelines for maintaining confidentiality and ethical practices was part of the research process (see Appendix C). To protect the confidentiality of the participants, a code replaced each participant's name. Secure storage of all study data and documents in a locked briefcase ensured confidentiality of the data obtained during the interviews. All study materials will be destroyed three years following completion of the study.

Data Collection

The current qualitative, phenomenological study involved collecting data through in-depth, unstructured interviews with open-ended questions. The participants were interviewed in person or on the telephone. Telephone interviews were appropriate because of the geographical dispersion of the participants in the Hawaiian Islands and in Florida. The face-to-face interviews were conducted in an office for which signed permission to use the premises was obtained.

During the interviews, the researcher, who has a master's degree in counseling, was able to establish a relaxed environment in which the participants felt comfortable sharing personal and sensitive information. The researcher employed listening skills and engagement techniques to encourage open and authentic communication throughout the interview process. Social cognitive approaches to meaningful dialogue, based on the theory that actions are usually rooted in thoughts concerning healthy lifestyle choices, helped capture the participants' views about trust in the medical community (McCormick & Martinko, 2004). Exploring cognitions related to self-regulation enabled a qualitative examination of the participants' thought processes.

The interview protocol included all the steps necessary in the interview process. The interviews started with a discussion of the informed consent form and request for the participants' signature on the form or by e-mail when participants were interviewed by telephone. After providing consent, the participants received a copy of the interview questions.

In conjunction with thanks for contributing to the study, each participant received an explanation of the purpose of the study and of the interview process. Next, each participant spent two minutes reviewing a definition of *genocide* and the interview questions. The interview began with the audio recorder turned on.

The focus of the interview questions (see Appendix D) was the participants' perceptions about the genocide theory and the Tuskegee experiments, and their views on whether either of those topics affected their sexual behavior. The interview questions addressed perceptions about HIV/AIDS and specific sexual behaviors, such as the use of protection during sex. After the participants answered all the interview questions, they received thanks for participating in the study, and the interview was concluded.

Data collection and analysis were completed concurrently to determine when data saturation was achieved. This technique is called *data analysis in action*. Ary et al (2006) explained that data analysis in action involves "reviewing the data as they are being collected and attempting to synthesize and make sense out of what is being observed" (pg. 640).

Member checking was part of the research process. Ary et al (2006) reported that member checking is a process in which the participants review the researcher's data and interpretations and indicate whether the participants agree with the findings. Member checking increases the credibility and validity of the research (Trochim, 2006).

Data Analysis

Analysis and interpretation took place concurrently during data collection to ensure data saturation (Ary et al, 2006). The subjective art of analyzing and interpreting phenomenological data involves carefully coding the information collected. In the current study, the data reflected the lived experiences and perceptions of African-Americans about the Tuskegee experiments, the genocide theory, and sexual health decisions. The data analysis process was inductive, from data to interpretation (Ary et al, 2006). Analyzing and interpreting the data represented the most important part of the research process to illuminate the problem under study. The coding, reducing, and interpreting process was followed.

In-depth, unstructured interviews were conducted with a purposive sample of 16 participants. Steps in the analysis included (a) rewriting the responses verbatim, (b) coding each participant's responses, and (c) grouping recurring patterns of codes into themes. Through a process of reduction, the 17 themes initially identified became five major themes.

The analysis included comparing the responses of men

and women. Another step consisted of paraphrasing the participants' responses in the context of the most salient themes. A table was constructed to illustrate each theme and included the participants' codes, age, and gender. The themes led to the development of recommendations and outcomes of the study.

Summary

Chapter 3 included an overview of the research methodology used in the current study. The chapter began with a discussion of the research method and design and their appropriateness for the study. The study population, sample, and sampling method were other topics addressed in the chapter, followed by the informed consent process, the measures taken to maintain the confidentiality of the participants, the data collection method, considerations regarding validity and reliability, and the data analysis process.

Using the qualitative, phenomenological approach and conducting one-on-one interviews with the participants resulted in in-depth information about the varied perspectives and lived experiences of African-Americans about the Tuskegee experiments, the genocide theory, and sexual health decisions. Analysis of the data resulted in a greater understanding of factors that influence African-Americans' sexual health decisions and, by extension, their likelihood of contracting HIV/AIDS. Chapter 4 contains a presentation of the study findings. Government leaders, health-care providers, and other stakeholders might use the findings of the study to develop ways to reduce the rate of HIV/AIDS in the African-American population.

4

PRESENTATION AND ANALYSIS OF DATA

THE PURPOSE OF this hermeneutic, phenomeno-logical research study was to explore the perceptions, understandings, opinions, and lived experiences of African-American men and women about conspiracy theories and sexual health decisions. The following two research questions guided the inquiry:

R1.   What are the lived experiences and per-ceptions of African-Americans pertaining to conspiracy beliefs, and the influence of such beliefs on their sexual behaviors?

R2.   What influence, if any, do conspiracy beliefs held by African-Americans have on their decisions associated with seeking medical attention?

Information stemming from this study may provide leaders and policy makers with insight for strategies to combat the higher epidemiological rates of HIV/AIDS among African-Americans.

Chapter 4 includes a presentation of the data obtained during interviews and findings that address the two research questions. An interview protocol included 11 questions. The chapter provides demographic information and relevant details of participant feedback. The participants either lived or had lived in various parts of the United States, including Hawaii. Nine women and seven men were recruited and interviewed for the study. All participants identified their race as African-American. The participants' ages ranged from 28 to over 60.

Synopsis of the Research Methodology

Based on the research questions developed for the study, the most suitable method of sampling was purposive qualitative sampling, consisting of choosing participants who could most effectively assist the researcher with understanding their ontological perspectives. Sampling methods that best fit a type of study produce abundant results, accurate outcomes, detailed responses, and useful insights to learn about a social phenomenon. Participants were asked to reflect on their knowledge of a known research project called the Tuskegee experiments, conducted in Tuskegee, Alabama, between 1932 and 1972. The following two chapters provide summaries of the research process and findings.

In-depth interviews took place with 16 recruited participants selected through a combination of snowballing and purposive sampling methods. Each participant engaged in a step-by-step process consisting of providing pertinent documentation, reading the informed consent and confidentiality agreements, and signing and dating each required document. Before starting their interview, the participants listened to instructions the researcher read from the interview protocol. Participants answered questions about their comfort level for noise and other interferences, as well as about issues potentially compromising confidentiality.

The classifications of data collection included (a) demographic data, (b) factor analysis, and (c) data recorded and transcribed from 16 willing and qualified participants who responded to 11 phenomenological questions. Steps included rewriting the interviews verbatim and coding each participant's responses. Bracketing led to groups of themes identified recurring patterns of codes. Through reduction, the initial 17 themes became five overarching themes. Analysis included comparing the responses from men and from women. Paraphrasing the participants' responses highlighted the most salient themes. The final step consisted of developing recommendations and identifying outcomes from the study, based on the selected overarching themes.

DATA COLLECTION

The qualitative, hermeneutic, phenomenological design best represented data collection and participant outcomes. Various components of data collection in a qualitative research study include observations, interviews, documents, and audio materials. This hermeneutic, phenomenological research study involved interviews with open-ended questions and questions sometimes requiring only one-word answers to address the two research questions.

Two crucial elements of data collection were the recording and transcription of the interviews into a computer file, using code names to protect the identity and confidentiality of the participants. Eleven interview questions facilitated obtaining narrative responses from the participants about their perceptions of (a) genocide conspiracy theories related to sexual health decisions and choices and (b) the potential influence of African-Americans' lack of trust in the medical community or individual doctors on decisions to seek treatment or screening. Technical difficulties during the interviews became a limitation when certain answers were inaudible. When this occurred, the term inaudible was used in the transcript.

Five interviews took place by telephone, and 11 interviews took place in person. The study did not include electronic mail interviews, because of confidentiality concerns. Codes with Greek letters were assigned to the participants and protected their identities. Participants' input was quoted to highlight recurring themes or to emphasize the extraordinary nature of their views. Qualitative hermeneutic research relies on the use of thick, rich, descriptive data, and generally uses extensive quotations to achieve the goal of identifying overarching themes. Quotations are used to highlight recurring themes or to emphasize the extraordinary nature of a participant's views.

INTERVIEW PROCESS

Initial contact with the potential participants occurred by telephone or the social media tool Facebook in private boxes. Conversation between the researcher and potential participants included (a) the research topic, (b) logistical barriers to overcome for confidential interviews, (c) the topic of HIV/AIDS in the African-American community, (d) protocol for confidential referrals, (e) confidentiality, (f) informed consent, (g) the interview protocol, and (h) publishing availability. The definition of a key term was read to participants in addition to instructions and the actual research questions.

Interview scheduling was a major challenge in that many of the participants lived on the mainland with vastly different time zones. Each participant received a copy of the interview protocol to read along with the questions. When signing the interview protocol, each participant agreed to have the responses tape-recorded for ease of transcribing.

Participants provided the required documentation to move forward with their interview, including signed and dated confidentiality agreement, informed consent, and interview protocol, in person or by U.S. mail. Only the interview protocol was e-mailed to a few research participants. To avoid compromising confidentiality, no faxed, scanned, or electronic mail correspondence requiring a response was allowed. Although Skype was an option for the participants, no Skype interviews took place.

Participant Demographics

Conscious efforts led to participant selection to obtain a sample of individuals with higher educational levels and individuals who had completed only high school. No high school dropouts participated in interviews for the study. Several threats to internal validity exist in qualitative research. In this study, threats to validity included (a) history (i.e., interview environment, time passing between the pilot study and the main study); (b) selection (i.e., significantly more of one gender recruited over the other, lopsided education levels, and randomness); (c) diffusion of treatments (i.e., when a married couple referred source participants, there was always potential to compare notes); (d) resentful demoralization (i.e., HIV/AIDS affects African-Americans more than any other ethnic group); and (e) instrumentation (when participants see the interview questions in advance, they have time to fabricate responses, an opportunity to *fake good*).

For this study, interaction of setting and treatment proved most vulnerable, in the sense that HIV/AIDS incidence occurs on varying levels from state to state. For instance, the percentage of HIV/AIDS overall in Hawaii is only 1% of the total population (CDC, 2009). The threat was mitigated with technology and the researcher's skill; a relationship was established with participants residing in other states through the use of well-documented clinical therapy techniques of active listening and reflection, enabling easy exchange of discourse.

Through snowballing and purposive strategies, 140 potential research participants were identified from across the United States and Hawaii. All participants identified through social media were eliminated from the study. Sixteen participants identified through snowballing and purposive strategies qualified for participation and were interviewed.

Five participants were married, two married participants had no children, five single participants had children, and four single participants had no children. Of the participants in interviews, six had educational levels spanning from high school to some college. Three participants held bachelor's degrees, four participants held master's degrees, and three participants held doctoral degrees.

The researcher tried to align participants in terms of various ranges of variables like age, education, and geographical area, to match the spectrum of individuals afflicted by HIV/AIDS. Having HIV or AIDS was not a condition for participating in this study. Figures 3 to 6 include graphic accounts of participant information and a recruitment profile pie chart. Recruitment for specific individuals was difficult because of the location of the researcher in Hawaii.

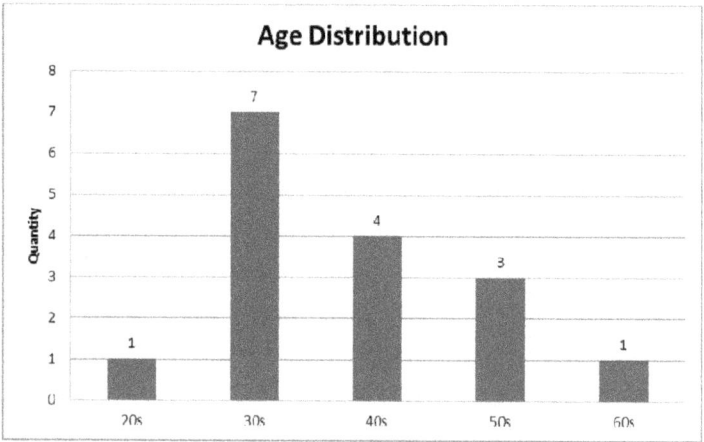

Figure 3. Age groupings of participants chosen for this study.

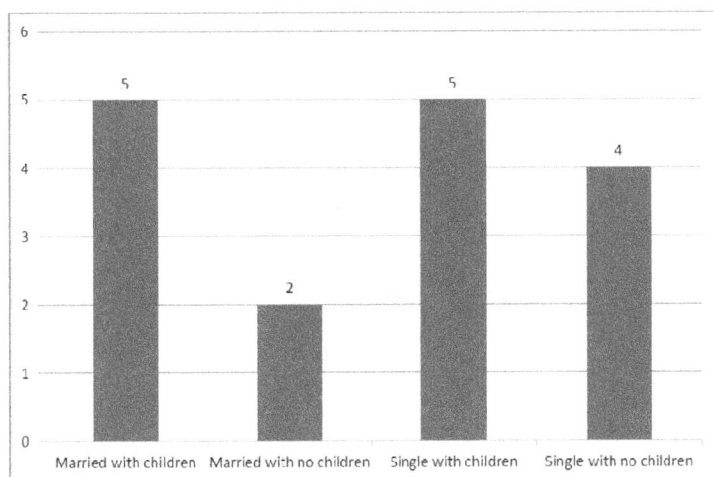

Figure 4. Participants' family status chosen for this study.

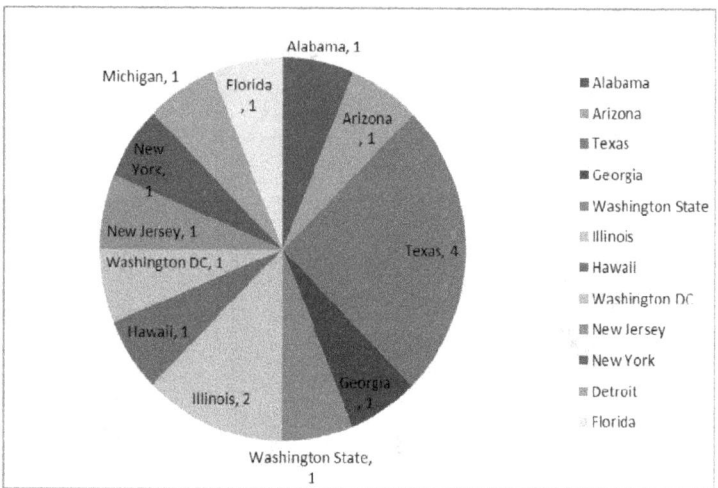

Figure 5. Geographic locations of participants chosen for this study.

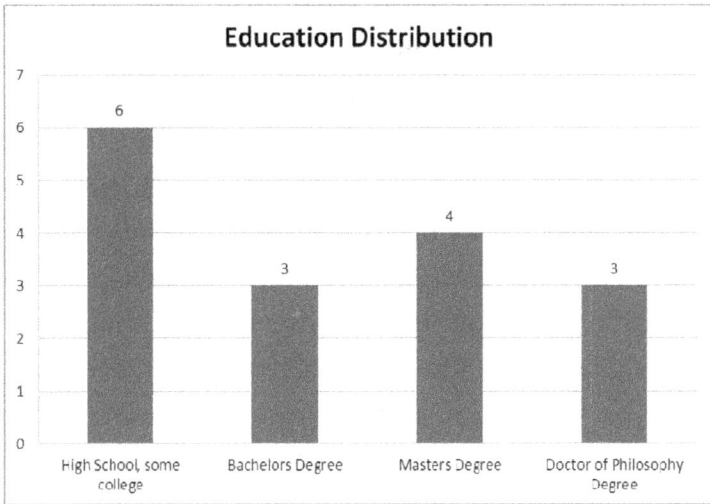

Figure 6. Educational levels of participants chosen for this study.

Data Analysis Review

The process of synthesizing and analyzing the material obtained during interviews includes becoming familiar with the data by reading, reflecting, describing, and classifying responses carefully. These functions served to prepare the researcher for identifying codes, themes, and overarching themes. Broader themes gave way to narrower ones, and the research concluded with findings that either supported or refuted the research questions.

The exercise of coding often leads to new questions. The researcher addressed the original research questions, provided information on gaps in research, and finalized the process with conclusions and recommendations. The intervention technique of active listening while interviewing was a process to test the research questions for potential changes. The next section includes the answers from the 16 participants in the interviews.

Data Analysis Outcomes

The research questions were as follows:

R1.   What are the lived experiences and perceptions of African-Americans pertaining to conspiracy beliefs and the influence of such beliefs on their sexual behaviors?

R2.   What influence, if any, do conspiracy beliefs held by African-Americans have on their decisions associated with seeking medical attention?

The 11 interview questions were as follows:

1. How do your perceptions on conspiracy of genocide influence your sexual health choices?
2. What are your perceptions regarding whether there was a conscious intent to create a virus capable of controlling certain populations, mainly minorities, and if so, how and why?
3. What is your knowledge of the Tuskegee experiments, and what are your thoughts on this topic as it relates to HIV/AIDS?
4. Do conspiracy beliefs of genocide influence your decision to bypass getting screened for HIV/AIDS in order to know your HIV status, and if so, how?
5. When deciding to engage in sex, do conspiracy beliefs related to genocide or the mass killing of African-Americans factor into your decision?
6. What are your perceptions of the threats African-American families face when someone they know is infected with HIV/AIDS?
7. What methods of protection do you think are important to continually practice?

8. What are your perceptions about why African-Americans are disproportionately affected by the disease HIV/AIDS?
9. What are your thoughts about condom use?
10. Do you use a condom (male or female) when engaging in sexual activity?
11. If you do not use a condom, why not?

Participant feedback on key points in these interview questions provided the basis for recommendations and outcomes for this research study. This document does not include full transcriptions of each participant's responses to protect the participants' confidentiality. Instead, quotations from each participant resulting from coding and bracketing represent the direction of the ultimate conclusion and recommendations discussed in Chapter 5.

INITIAL THEMES

The first step in analysis produced the following 17 themes from coded data:

1. Conspiracy of genocide does not influence sexual health choices.
2. Conspiracy of genocide does influence sexual health choices.
3. There was no conscious intent to create a virus capable of controlling certain populations, mainly minorities.

4. AIDS is a virus that American scientists created to experiment with African-Americans, and it spread out of control.

5. Many black men are not dating African-American women and are passing HIV/AIDS to non-black populations.

6. Christian men harbor ambivalent perspectives and values on sex (perhaps sometimes unconsciously), entering consenting intimate relationships unprepared while leaving women vulnerable.

7. Down-low brothers (men who have sex with men, or MSM) become infected and pass HIV/AIDS to other races.

8. American scientists at Tuskegee Institute were morally wrong for their study on the effects of the syphilis virus on African-American males.

9. African-Americans feel stereotyped when considering HIV/AIDS.

10. There exists a conspiracy against black women who are being unfairly targeted for HIV/AIDS and are being asked to be tested.

11. There is a lack of trust for doctors among African-Americans.

12. Threats include ostracism, ambivalence as to where to go for care and treatment, financial burden on the family, that other family members, if exposed to the infected person, could also catch HIV; the person may be perceived as being *gay*, and most

frequently cited was a lack of health care, having to listen to the doctor and pressure to follow certain medical protocols. African-Americans who only mate with other African-Americans are living in densely populated areas where the disease has spread.

13. Methods of protection commonly used are (a) condoms and spermicide, (b) abstinence, and (c) refraining from having more than one committed sexual partner.

14. African-Americans are more disproportionately affected by HIV/AIDS, largely due to ignorance and lack of education about how the virus is spread and which methods of protection are most effective. The ignorance stems from lack of knowledge early enough so that people are not infecting others perhaps by the thousands.

15. People testing positive for HIV do not believe they are actually positive and harbor a belief that HIV/AIDS medication actually causes AIDS.

16. Heterosexuals who do not know their status are engaging in unprotected sexual activity with other heterosexuals and contracting and passing along the HIV virus unknowingly.

17. Doctors cannot be trusted.

HERMENEUTIC INTERPRETATION OF PARTICIPANT COMMENTS

The researcher reduced the original 17 themes to five themes, using a filtering process. Feedback on conspiracy belief and perceptions of conscious intent were filtered as follows:

Participant Beta. The following are the results:

Position 1. Conspiracy beliefs of genocide influencing sexual health choices:

1. "There's been a conspiracy since the beginning of time."
2. "Conspiracy of genocide is real."
3. "I have been a victim of this 'conspiracy."
4. "Doctor gave me a C-section."
5. "Conspiracy of genocide affected my choice in the number of children I had versus how many I wanted."

Hermeneutic translation: Participant Beta perceives genocide exists.

Position 2. Conscious intent to create a virus:

1. "Yes, I believe there was conscious intent to create a virus capable of controlling blacks."

2. "It was more than intention; the conspiracy actually happened."

Hermeneutic translation: Participant Beta perceives there was conscious intent to create a virus.

Participant Gamma. The following are the results:

Position 1. Conspiracy beliefs of genocide influencing sexual health choices:

1. No belief in conspiracy of genocide.
2. Does not affect sexual health choices.

Hermeneutic translation: Participant Gamma does not have conspiracy beliefs about genocide.

Position 2. Conscious intent to create a virus:

1. "No, there was no conscious intent for this; AIDS appeared in the '80s. HIV/AIDS was not intentionally introduced."

Hermeneutic translation: Participant Gamma perceives no conscious intent occurred to create a virus, and HIV/AIDS began and spread in the 20th century.

Participant Epsilon. The following are the results:

Position 1. Conspiracy beliefs of genocide influencing sexual health choices:

1. "At the onset of it, did not think of HIV in terms of genocide."
2. "Didn't think of conspiracy of genocide in the same thoughts of sexual health choices."
3. "Fear of HIV/AIDS has affected my sexual health choices."
4. "Ideas and thoughts of conspiracy of genocide do affect my sexual health choices today."

Hermeneutic translation: Participant Epsilon perceives, initially, that HIV/AIDS may have been incidental, but now believes evidence is pointing to factors that could prove otherwise.

Position 2. Conscious intent to create a virus:

1. "Evidence has convinced me there may be a conspiracy theory."
2. "HIV in Africa resulting from monkeys is evidence there could be a conspiracy."
3. "Ideas of conquering people whose immune systems cannot fight off viruses."

4. "Evidence points to a possible conspiracy theory supporting possible conscious intent to create a virus capable of controlling certain populations."

Hermeneutic translation: Participant Epsilon perceives theories pointing to HIV/AIDS in monkeys in Africa support the concept of a possible conspiracy.

Participant Theta. The following are the results:

Position 1. Conspiracy beliefs of genocide influencing sexual health choices:

1. "Don't consider conspiracy beliefs."
2. "Think about conspiracy of genocide when considering sex health choices but undecided on whether it is real."

Hermeneutic translation: Participant Theta does not consider conspiracy beliefs.

Position 2. Conscious intent to create a virus:

1. "There may or may not have been a conscious intent to create a virus capable of controlling certain populations."
2. "I've heard the theory. Undecided on it as being true/real."

Hermeneutic translation: Participant Theta perceives there may have been conscious intent to create a virus, has heard of conspiracy theories, but is undecided over whether they are real or not.

Participant Eta. The following are the results:

Position 1. Conspiracy beliefs of genocide influencing sexual health choices:

1. "I really haven't given too much thought on. I believe that a person should take care of themselves if they are with their personal hygiene's . . . and realize the importance of understanding what might be a conspiracy."
2. "Believes monogamy is the way to go."
3. "Protect yourself from STDs."

Hermeneutic translation: Participant Eta does not have perceptions on the subject of conspiracy of genocide in relation to sex health choices, does not think of genocide in relation to sexual health choices, and asserts individuals should be responsible for their own sex health care.

Position 2. Conscious intent to create a virus:

1. "Not sure if this is a valid theory."

2. "I believe it has been done before."
3. "A virus may have been created to control a race."

Hermeneutic translation: Participant Eta believes a virus may have been created to cause harm or to control a race.

Participant Iota. The following are the results:

Position 1. Conspiracy beliefs of genocide influencing sexual health choices:

1. "Conspiracy of genocide does not factor."
2. "There is no conspiracy of genocide-constructed HIV/AIDS."
3. "There may have been a virus created."

Hermeneutic translation: Participant Iota believes there may have been a virus created, but conspiracy of genocide does not factor in when making health choices.

Position 2. Conscious intent to create a virus:

1. "The motive may have been to keep minorities out of power."
2. "They probably achieved keeping blacks from power by creating a disease transmittable through unprotected sexual activities—since sex is a need like air, water, and food."

3. "The strategy may have been to (inaudible) the dwindling of African-American population over time by agent HIV/AIDS."

Hermeneutic translation: Participant Iota perceives there may have been a virus created with a motive to obliterate African-Americans and keep them out of power.

Participant Lambda. The following are the results:

Position 1. Conspiracy beliefs of genocide influencing sexual health choices:

1. "Abstinence, condoms, and choosing sex partners carefully are important to prevent the spread of HIV/AIDS."

Hermeneutic translation: Participant Lambda has become more cautious about sexual health choices/decisions.

Position 2. Conscious intent to create a virus:

1. "I am one who believes viruses were created to control population, and they grew out of control."
2. "The human population is too big for this earth."

Hermeneutic translation: Participant Lambda perceives that a virus or viruses were created with conscious intent to harm groups of people in an effort to control the growth of human population.

Participant Mu. The following are the results for Participant Mu:

Position 1. Conspiracy beliefs of genocide influencing sexual health choices:

1. "There was no theory on conspiracy of genocide in reality."

Hermeneutic translation: Participant Mu does not believe in a conspiracy.

Position 2. Conscious intent to create a virus:

1. "Genocide occurred against Jews and blacks through slavery, but the concept differs because to engage in sex is a choice."

Hermeneutic translation: Participant Mu perceives that genocide has occurred in American history; however, HIV/AIDS requires one to make conscious choices over whether to protect oneself and others from venereal diseases.

Participant Nu. The following are the results:

Position 1. Conspiracy beliefs of genocide influencing sexual health choices:

1. "African-Americans are not educated about STDs and need to engage personal responsibility to access health care and education. These will lead to better sexual choices."

Hermeneutic translation: Participant Nu does not believe in the existence of a conspiracy, but believes that African-Americans are not being responsible about sexual health choices and practices and lack the necessary education about HIV/AIDS.

Position 2. Conscious intent to create a virus:

1. "HIV/AIDS virus was not created to harm African-Americans."
2. "Better education will lead to better sex health choices."
3. "There was no conscious intent to create a virus (HIV/AIDS) to harm African- Americans."

Hermeneutic translation: Participant Nu does not believe that HIV/AIDS was created and believes there was no conscious intent to create a virus that causes harm;

however, lack of education among African-Americans is prohibiting individuals from protecting themselves.

Participant Xi. The following are the results:

Position 1. Conspiracy beliefs of genocide influencing sexual health choices:

1. "Conspiracy of genocide affects my sexual choices."
2. "Cause of diseases are man-made and are put here as a weapon to destroy us using a behavior that is a human need."

Hermeneutic translation: Participant Xi perceives HIV/AIDS and other diseases are man- made, and conspiracy beliefs impact participant Xi's sexual health choices.

Position 2. Conscious intent to create a virus:

1. Asians and Indians are the largest minority group, not a threat to humanity (fear) or the world. (Color the world threat.) African-Americans are a threat to whites. HIV/AIDS was introduced to the African-American communities to control the population. Control a race . . . keeps power. (Participant Xi)

Hermeneutic translation: Participant Xi perceives whites feel threatened by people of color, and the atrocities perpetrated against African-Americans were conscious, with the intent to prevent an overwhelming population of people of color from existing in order to maintain leverages of power.

Participant Omicron. The following are the results:

Position 1. Conspiracy beliefs of genocide influencing sexual health choices:

1. "Conspiracy of genocide does not influence my sexual health choices."

Hermeneutic translation: Participant Omicron believes that conspiracy of genocide has no influence on sexual health choices.

Position 2. Conscious intent to create a virus:

1. "I don't believe that people made conscious intent of creating a virus capable of controlling certain populations."

Hermeneutic translation: Participant Omicron does not believe there were conscious actions to create a virus for malicious intent.

Participant Pi. The following are the results:

Position 1. Conspiracy beliefs of genocide influencing sexual health choices:

1. "I do not have thoughts about genocide and sexual health choices."

Hermeneutic translation: Participant Pi does not make the connection between conspiracy belief and sexual health choices.

Position 2. Conscious intent to create a virus:

1. "I do not believe a disease was created to bring down a community."
2. "Human experiments have been taking place that allow a germ to spread and fester."
3. "Something has been spread in order to wipe out populations throughout history."
4. "Native Americans have been the victims of such attacks. White settlers were guilty of such atrocities."

Hermeneutic translation: Participant Pi points out evidence of a history of medical ethics breaches that have harmed American Indians and African-Americans; however, despite the evidence, Participant Pi does not perceive diseases were created to "bring down a community."

Participant Rho. The following are the results:

Position 1. Conspiracy beliefs of genocide influencing sexual health choices:

1. "Mothers should get involved with educating sons about sex and STDs."
2. "Make the connection between unprotected sex and venereal disease."

Hermeneutic translation: Participant Rho believes sexual health choices stem from what is taught at home.

Position 2. Conscious intent to create a virus:

1. "History tells you Tuskegee experiments took place to possibly control a population."
2. "The Mid-Atlantic transfer is evidence of a historical event that set the stage for Tuskegee experiments."
3. "There is other history to suggest that conscious intent was possible."
4. "Current trends with health disparities and death rates in America shows conscious intent was or is possible."
5. "Genocide by another means, in another context, could be represented today by health disparities, poverty, and injustice."

Hermeneutic translation: Participant Rho provides historical perspective on the Tuskegee experiments and the Mid-Atlantic transfer to illustrate the basis for the perception that genocide by another means, in another context, could have been represented by the atrocities perpetrated against blacks and that other trends such as health disparities and possible chemicals in foods are disease-causing, maybe influencing actions to cause fatal harm. (Participant Rho)

Participant Tau. The following are the results for Participant Tau:

Position 1. Conspiracy beliefs of genocide influencing sexual health choices:

1. "Thoughts of conspiracy of genocide make me not want to consummate with black men."

Hermeneutic translation: Participant Tau sheds light on personal trust factors for black men in conjunction with the perception of conspiracy belief.

Position 2. Conscious intent to create a virus:

1. "Race wouldn't have mattered. Black men were convenient targets."
2. "Scientists were aware of the ill effects of syphilis."
3. "Unethical behavior was perpetrated against black

men. Black men at the time were a convenient target. This could have happened to any group."

Hermeneutic translation: Participant Tau perceives there were historic foundations that support a theory for conspiracy to create a virus with intent to cause serious harm.

Participant Upsilon. The following are the results:

Position 1. Conspiracy beliefs of genocide influencing sexual health choices:

1. "They do not affect my sexual health choices."

Hermeneutic translation: Participant Upsilon perceives conspiracy beliefs do not affect Participant Upsilon's sexual health choices.

Position 2. Conscious intent to create a virus:

1. "Ummm, I wouldn't say that I (pause) don't believe so . . . . I guess that I, I, I, I guess that I am familiar with a certain, I guess historical uh, I guess facts of uh of uh I don't know if it was a virus but certain diseases that were being spread in the community."
2. "In my mind the why part would be to eliminate those populations or to severely affect their growth."

Hermeneutic translation: Participant Upsilon believes a virus was "being spread" and with the intention to "eliminate" certain populations or contain their growth.

Participant Chi. The following are the results for Participant Chi:

Position 1. Conspiracy beliefs of genocide influencing sexual health choices:

1. "Protect oneself from genocidal intent."
2. "AIDS in the mid-1980s affected the homosexual population initially."
3. "Don't engage in multiple sex partners."

Hermeneutic translation: Participant Chi recalls the AIDS epidemic of the 1980s and perceived at the time that homosexual behavior caused the epidemic, not a conspiracy.

Position 2. Conscious intent to create a virus:

1. "There may have been an attempt to wipe out the African population through AIDS."
2. "There may have been an attempt to control the growth of the African population."
3. "There could be an unpopular perception that the minorities are becoming the new majority."

4. "When you look at the numbers, it's showing that it's affecting the black community."

Hermeneutic translation: Participant Chi perceives there was a conspiracy to cause harm, but individuals need to be responsible about personal choices to protect themselves and others.

The next section provides details of the five overarching themes extracted from the responses to the 11 interview questions. The sensitive nature of the research questions prohibited using verbatim transcripts. To protect the identity and confidentiality of willing, qualified participants, the next section includes only excerpts supporting the themes extrapolated from the codes in participants' responses.

Overarching Themes

In this hermeneutic, phenomenological research study, repeated coding of the participants' responses produced 17 themes, from which five overarching themes were identified (see Table 2). The 11 interview questions were structured to obtain the participants' perspectives about conspiracy theory and their attitudes toward doctors or the medical community in America.

The participants did not have time to examine the 11 interview questions and prepare responses in advance.

Bracketing and coding of the data obtained from the 11 interview questions produced five overarching themes. Table 2 shows the cumulated responses by percentage.

Female perspectives from Questions # 1, 2, 4, 6, and 8. Below are quotes supporting the five overarching themes from each female participant.

Participant Beta. Participant Beta stated the following (numbers indicate the interview questions):

Q1. How do conspiracy beliefs of genocide influence your sexual health choices? "There have been a conspiracy ever since the beginning of time, as far as back as I can remember, to limit as many black babies from being born as possible."

Q2. What are your perceptions regarding whether there was a conscious intent to create a virus capable of controlling certain populations, mainly minorities, and if so, how and why? "Yes, I believe there was a conscious intent to create a virus capable of controlling the black population . . . AIDS was something that was created by scientists."

Q4. Do conspiracy beliefs of genocide influence your decision to bypass getting screened for HIV/AIDS in order to know your HIV status, and if so, how? "Yes . . . as I don't feel I have a reason to feel like I need to be test, due to my conspiracy belief of genocide, my decision is highly affected. *'Cause I don't trust doctors.*"

Q6. What are your perceptions of the threats African-American families face when someone they know is infected with HIV/AIDS? "Contracting that disease." "Lack of knowledge about the infection." "Fear—afraid of all the bodily fluids . . . we should educate ourselves."

Q8. What are your perceptions about why African-Americans are disproportionately affected by the disease HIV/AIDS? "I have no proof to believe black people are dis—that African-Americans are disproportionately affected by the disease HIV and AIDS . . . I don't personally believe that we are disproportionately affected."

Participant Gamma. Participant Gamma stated the following (numbers indicate the interview questions):

Q1. How do conspiracy beliefs of genocide influence your sexual health choices? "I actually do not have any beliefs that there is a connection between uh genocide and sexual health practices."

Q2. What are your perceptions regarding whether there was a conscious intent to create a virus capable of controlling certain populations, mainly minorities, and if so, how and why?

> "I do not believe there is a conscious intent to create a virus to control certain populations. . . I'm well aware of experiences and experiments that have done such, but with this particular disease I don't think it was something that was intentional." (Participant Gamma)

Q4. Do conspiracy beliefs of genocide influence your decision to bypass getting screened for HIV/AIDS in order to know your HIV status, and if so, how? "Conspiracy beliefs do not affect my decision to bypass getting screened for HIV/AIDS."

Q6. What are your perceptions of the threats African-American families face when someone they know is infected with HIV/AIDS? "Better health education on this topic would really help out uh this demographic."

> "African-American family they may face issues of shame . . . issues of guilt . . . a lot of internal uh reflection . . . I think they face financial burdens, I think they face social disassociation, the family may want to shun that person . . . there are some myths that are associated with HIV/AIDS that are associated with African-Americans that may cause them to react in negative ways towards individuals who have HIV/AIDS rather than positive ways." (Participant Gamma)

Q8. What are your perceptions about why African-Americans remain disproportionately affected by the disease HIV/AIDS? "They don't know their status; there is an internal fear of getting the information . . . because the fear is they'd get bad results."

Participant Epsilon. Participant Epsilon stated the following (numbers indicate the interview questions):

Q1. How do conspiracy beliefs of genocide influence your sexual health choices? "The fear of HIV/AIDS has affected my choices, you know? But I didn't think about it in the terms of genocide, in the beginning."

Q2. What are your perceptions regarding whether there was a conscious intent to create a virus capable of controlling certain populations, mainly minorities, and if so, how and why? "Initially I never gave it a thought 'til um . . . evidence started creeping up."

Q4. Do conspiracy beliefs of genocide influence your decision to bypass getting screened for HIV/AIDS in order to know your HIV status, and if so, how? "No, um. It hasn't. What has actually affected my decision . . . . It was actually a fear of being told that I would have it."

Q6. What are your perceptions of the threats African-American families face when someone they know is infected with HIV/AIDS? "The main threat they face is the ability to able to pay for treatment." "Engaging in sex by people who are uh um, known to be gay and African-Americans as a whole don't accept people—especially in the church—who are gay."

> "The biggest threat is sometimes being found out that they are not heterosexual . . . engaging in homosexual sex acts or, or, or the down-low thing being on the DL and

also then engaging in heterosexual um relationships and then transmitting the AIDS
virus the HIV virus to a partner unknowingly." (Participant Epsilon)

Q8. What are your perceptions about why African-
Americans remain disproportionately affected by the disease HIV/AIDS?

"My main theory is that um, a lot of
African-American males like to tell females
they don't enjoy condoms 'cause they don't
feel anything . . . another really, really, really
big thing with this is that um we sometimes
don't believe that, you know, it can happen
to us." (Participant Epsilon)

Participant Theta. Participant Theta stated the following (numbers indicate the interview questions):
Q1. How do conspiracy beliefs of genocide influence
your sexual health choices? "I don't really consider this in
my sexual health choices."
Q2. What are your perceptions regarding whether
there was a conscious intent to create a virus capable of
controlling certain populations, mainly minorities, and if
so, how and why? "I don't have a strong opinion about it
um. I don't believe it firmly. I also don't deny its possibility."
Q4. Do conspiracy beliefs of genocide influence your

decision to bypass getting screened for HIV/AIDS in order to know your HIV status, and if so, how? "They do
not."

Q6. What are your perceptions of the threats African-
American families face when someone they know is infected with HIV/AIDS? "The same threats that others, um
other races with families, family members that are infected
with HIV. Simply because we are aware of how the virus
is transmitted."

Q8. What are your perceptions about why African-
Americans are disproportionately affected by the disease
HIV/AIDS? " I don't know the answer to that question."

Participant Lambda. Participant Lambda stated the
following (numbers indicate the interview questions):

Q1. How do conspiracy beliefs of genocide influence
your sexual health choices? "I'm more cautious. I'm more
cautious in choosing my partners and using protection."

Q2. What are your perceptions regarding whether
there was a conscious intent to create a virus capable of
controlling certain populations, mainly minorities, and if
so, how and why?

> "Yes, I do believe that there were vi-
> ruses that were created . . . the HIV virus
> it was created, and it just got away . . . I
> do believe there were viruses created, why,
> because we're getting too big for this earth."
> (Participant Lambda)

Q4. Do conspiracy beliefs of genocide influence your decision to bypass getting screened for HIV/AIDS in order to know your HIV status, and if so, how? "No, I get tested."

Q6. What are your perceptions of the threats African-American families face when someone they know is infected with HIV/AIDS? "There's a possibility that they could get infected."

Q8. What are your perceptions about why African-Americans remain disproportionately affected by the disease HIV/AIDS? "I think because uhhhm, in some of the rural areas where African-Americans um reside there's a um lower income, less health care, um you know the awareness um isn't as prevalent."

Participant Mu. Participant Mu stated the following (numbers indicate the interview questions):

Q1. How do conspiracy beliefs of genocide influence your sexual health choices? "They don't, 'cause I don't believe that there is a conspiracy."

Q2. What are your perceptions regarding the belief that there may have been a conscious intent to create a virus capable of controlling certain populations, mainly minorities, and if so, how and why? "No, I do not."

Q4. Do conspiracy beliefs of genocide influence your decision to bypass getting screened for HIV/AIDS in order to know your HIV status, and if so, how? Inaudible response—faulty tape.

Q6. What are your perceptions of the threats African-American families face when someone they know is infected with HIV/AIDS? "Isolated and people not show-ing compassion.

. . the breakdown of family as it relates to no commu-nication, being ostracized, and looked upon as something dirty."

Q8. What are your perceptions about why African-Americans remain disproportionately affected by the disease HIV/AIDS? "We are still lacking like I said as it relates to knowing our status."

Participant Pi. Participant Pi stated the following (numbers indicate the interview questions):

Q1. How do conspiracy beliefs of genocide influence your sexual health choices? "I rarely think about genocide, and I really don't think about the two together."

Q2. What are your perceptions regarding whether there was a conscious intent to create a virus capable of controlling certain populations, mainly minorities, and if so, how and why?

> "I do or rather I don't necessarily believe that this occurred, that something has been created in order to, you know, to wipe out a population. But maybe something has been spread in order to wipe out populations, um; throughout history that's kind of been the case." (Participant Pi)

Q4. Do conspiracy beliefs of genocide influence your decision to bypass getting screened for HIV/AIDS in order to know your HIV status, and if so, how?

> "I have an awareness of history and what has happened, I do believe. I guess I am in a place where I feel more trusting of the current medical community, and so I don't feel if I go to get tested for HIV/AIDS that they are in any way going to be misleading—or, you know, lie about the results or anything like that." (Participant Pi)

Q6. What are your perceptions of the threats African-American families face when someone they know is infected with HIV/AIDS?

> "Their belief system as well as their awareness. They may not have health care. So sometimes we may not have access to health insurance, knowledge kind of in two ways, one . . . perhaps not knowing the person actually has HIV/AIDS maybe not getting tested so there's a barrier just right there of knowledge, um—but assuming that they do know that they have it . . . then, there can also be the barrier of knowing the proper channels to get help . . . sometimes they

can be believing so much in the church and
believing so much in the church only to the
point that um the belief in that prayer and
not doing something else like going to get
medication." (Participant Pi)

Q8. What are your perceptions about why African-
Americans remain disproportionately affected by the disease HIV/AIDS?

"Socioeconomic status . . . not having the
regular health-care checks and check-ups. . .
lack of knowledge . . . without protection it's
even a higher likelihood that the disease will
be spread . . . lack of maybe resources . . . lack
of knowledge early enough, you know, so that
by the time you find out you've already infected three or four people and/or they've already
infected three or four people and you don't
even know who to notify of or who to go back
and tell . . . when the Church is involved . . .
that um there's a tendency where people kind
of walk this thin line between moral right and
wrong and that conflict I think sometimes
causes people sort of to act one way but aspire
to something else . . . . Behaviors and desires
are in conflict so that they don't truly face the
actions that they are engaging in so they don't

enter it with preparations. It's kinda like sometimes like a man who might cheat on his wife.
He doesn't carry around condoms because he
doesn't want to seem like it's premeditated."
(Participant Pi)

Participant Tau. Participant Tau stated the following
(numbers indicate the interview questions):

Q1. How do conspiracy beliefs of genocide influence
your sexual health choices? "It makes me weary that if there's
been a genocide then I do not, especially if it is against a black
males . . . then I won't have sex openly with a black male."

Q2. What are your perceptions regarding whether
there was a conscious intent to create a virus capable of
controlling certain populations, mainly minorities, and if
so, how and why?

"They realized that a group of these
men were suffering from some anti, you
know, immune disease and it was easier to
just watch what happened to those men.
And they just happened to belong to a
certain ethnic group. Investigating. And it
didn't matter if they were black or white or
Latino." (Participant Tau)

Q4. Do conspiracy beliefs of genocide influence your
decision to bypass getting screened for HIV/AIDS in

order to know your HIV status, and if so, how? "I have been screened for HIV and AIDS simply because I was dating someone that had it."

Q6. What are your perceptions of the threats African-American families face when someone they know is infected with HIV/AIDS? "They are ostracized by society because the disease is seen more as a relationship to gay males or a relationship to Africans."

Q8. What are your perceptions about why African-Americans remain disproportionately affected by the disease HIV/AIDS? "Because, and then this is a female thinking about the male's standpoint, males . . . do not like to wear a condoms . . . . African-American males just do not like to wear a condoms."

Participant Upsilon. Participant Upsilon stated the following (numbers indicate the interview questions):

Q1. How do conspiracy beliefs of genocide influence your sexual health choices? "They do not affect my sexual health choices."

Q2. What are your perceptions regarding whether there was a conscious intent to create a virus capable of controlling certain populations, mainly minorities, and if so, how and why?

> "I guess historical uh, I guess facts of uh of uh I don't know if it was a virus but certain diseases that were being spread in the community. In my mind the why part

would be to eliminate those populations or to severely affect their growth."

Q4. Do conspiracy beliefs of genocide influence your decision to bypass getting screened for HIV/AIDS in order to know your HIV status, and if so, how? "No. Basically beliefs of genocide do not affect my decisions to get screened for HIV/AIDS."

Q6. What are your perceptions of the threats African-American families face when someone they know is infected with HIV/AIDS? "Lot of pressure to conform to the typical treatment and protocols for the infected person."

Q8. What are your perceptions about why African-Americans remain disproportionately affected by the disease HIV/AIDS?

"Areas that have a high population of people that are infected, um . . . there is a lack of education about how it's (HIV/AIDS) spread and how important it is to protect yourself .. . there may be substance use . . . . It's sort of not spread out in a sense . . . in one area it's harder to contain it's harder to prevent from spreading . . . lack of attention to it lack of, you know, funding among the people uuuuhm that are continually overlooked." (Participant Upsilon)

Summary of female perspectives. For Question #1, among female participants, five of nine women (56%) held no perceptions of whether there was or is a conspiracy of genocide, or did not think about conspiracy of genocide when considering sexual health choices. The other four women (44%) alleged there was or is a conspiracy with a genocidal theme that influenced their choices for decisions such as how many babies they could have, choosing partners, having sex or not with black men, and being more cautious when having sexual intercourse.

For Question #2, among women, four of nine participants (44%) perceived there was conscious intent to create a virus capable of controlling a population, namely minorities, through scientific means. One female participant cited evidence from past historic events, another female participant alleged that HIV/AIDS was an example of a virus that was intentionally created, and another female participant perceived that the intent may not have been to actually "wipe out a population" (Participant Pi). The remaining 22% of female participants, or two of nine, did not perceive a conscious intent to create a virus capable of controlling certain populations of people, namely minorities. Among female participants, one of the two did not feel that a virus could have or had been created at all.

Among female participants, three overarching themes emerged from answers to Question #4. The themes are (a) belief, or lack of, a conspiracy existed; (b) screens, or lack of, for HIV/AIDS; and (c) trust or lack of trust in the medical community. Only one of nine women (11%) reported lack of trust in medical professionals, namely doctors (Participant Beta), citing personal experiences to support her stance.

Participant Pi expressed having trust, an opposite view, in current medical doctors' ethical practices. Three of nine female participants (33.3%) reported they have screened for HIV/AIDS. The remaining women either did not outwardly state that they have either screened or have not screened for HIV/AIDS. Seventy-eight percent of female participants perceived there was no lack of trust in medical professionals, and one of nine (11%) felt extremely strongly that conspiracy belief of genocide and lack of trust for doctors influenced her sexual health choices (Participant Beta). The remaining 11% provided an inaudible response.

Among women, participant responses suggested the most salient issues on threats faced by African-American families was fear of bodily fluids, contracting HIV/AIDS, and lack of knowledge. Another female participant perceived such threats as shame, guilt, myths, and lack of health education about HIV/AIDS and how it spreads. Another female participant felt that one of the biggest threats African-American families face was heterosexuals engaging in sexual activity with other heterosexuals and contracting and passing HIV/AIDS unknowingly.

Another interesting theme arising from participant

responses among women was that the down-low culture (MSM—men who have sex with men) perpetuates HIV/ AIDS and the notion that women are not taking charge of their sexual health in efforts to please men. One could speculate that women who depend on men in the African-American community could be under pressure to perform without protection, whether it was real or perceived pressure.

Participant Epsilon alleged:

> "African Americans as a whole don't accept people—especially in the church—that are gay. The biggest threat was sometimes being found out that they are not heterosexual . . . engaging in homosexual sex acts or . . . or . . . or the down-low thing being on the 'DL' and also then engaging in heterosexual um relationships and then transmitting the AIDS virus the HIV virus to a partner unknowingly." (Participant Epsilon)

The comments may suggest that certain ethnic and religious cultures' attitude toward gay people poses as a barrier to gays who are not open to making wise sexual health choices.

Male perspectives. Below are quotations from each male participant regarding the overarching themes.

Participant Eta. Participant Eta stated the following (numbers indicate the interview questions):

Q1. How do conspiracy beliefs of genocide influence your sexual health choices? "Realize the importance of understanding what might be a conspiracy . . . basically it could be happening until uh it affects you . . . every person should be able to understand why they should protect their own um sexual health." (Participant Eta)

Q2. What are your perceptions regarding whether there was a conscious intent to create a virus capable of controlling certain populations, mainly minorities, and if so, how and why? "As far as validating I'm not really sure—but I do have my, uh, ideas regarding how the government does . . . and I would believe that it has been some things done in order to control populations."

Q4. Do conspiracy beliefs of genocide influence your decision to bypass getting screened for HIV/AIDS in order to know your HIV status, and if so, how? "The conspiracy belief it doesn't affect me that much to be honest with you. I hadn't really been tested that much. I'm not really an active individual."

Q6. What are your perceptions of the threats African-American families face when someone they know is infected with HIV/AIDS?

"People become very uh, uh vicious
in their attack of the family if they found
out (a person is infected with HIV/AIDS)

whether it's been in the paper or they say well what the person been . . . is just ostracizing often uh basically the alienation from the rest of the family that might occur."

Q8. What are your perceptions about why African-Americans remain disproportionately affected by the disease HIV/AIDS?

"Because we don't really have the funds that we need to get um, the necessary treatment . . . we don't really want no one to know what we do privately. Questions don't get answered a whole lot of times because they really didn't know the person. Lot of times it's been a kind a, a taboo or shun against because we being more aware of the consequences we just don't want the other family to be exposed or to have to go through the ridicule." (Participant Eta)

Participant Iota. Participant Iota stated the following (numbers indicate the interview questions):

Q1. How do conspiracy beliefs of genocide impact your sexual health choices?

"I would say it does not affect my thought process as far as a conspiracy. My interpretation on genocide is the killing of a race of people and in that aspect as far as conspiracy as far as my sexual health choices um I do not think at this time it plays a part in my life." (Participant Iota)

Q2. What are your perceptions regarding whether there was a conscious intent to create a virus capable of controlling certain populations, mainly minorities, and if so, how and why?

"Ummm, yes. I do believe there was a virus that might've been created . . . ummm and to control minorities so that um, the minorities cannot come in power and join together . . . people in power keep the power. And by doing that they eliminate . . . ummmm, by using basically, your own sex choices, they develop a virus . . . that uh, if you do not use protection, basically over time it will wipe out a race that might be um, more prone to unsafe sex practices." (Participant Iota)

Q6. What are your perceptions of the threats African-American families face when someone they know is infected with HIV/AIDS? "Possibility of being ostracized from your family."

Q4. Do conspiracy beliefs of genocide impact your

decision to bypass getting screened for HIV/AIDS in order to know your HIV status, and if so, how? "No um. I believe in protecting myself and those that uh, I may have sexual contact with so I definitely . . . get checked, get screened. I like to know what's going on."

Q8. What are your perceptions about why African-Americans remain disproportionately affected by the disease HIV/AIDS?

> "Rubbers uh, they take away the feeling, the sensation of enjoyment umm . . . but basically you know—we as African-Americans sometimes think we're immune to such things as those kinds of diseases . . . we used to think that it could only happen to gay people." (Participant Iota)

Participant Nu. Participant Nu stated the following (numbers indicate the interview questions):

Q1. How do conspiracy beliefs of genocide impact your sexual health choices?

> "I look at it more as personal responsibility . . . . You know and I feel that, that, that there is a conspiracy theory. I don't look at it as genocide. We not educated. The more educated we will be better, we will make better sexual choices." (Participant Nu)

Q2. What are your perceptions regarding whether there was a conscious intent to create a virus capable of controlling certain populations, mainly minorities, and if so, how and why? "No, I don't know as far as HIV and AIDS . . . smallpox . . . it was already a disease so . . . they used that against the Indians to destroy their race."

Q4. Do conspiracy beliefs of genocide impact your decision to bypass getting screened for HIV/AIDS in order to know your HIV status, and if so, how? "Uh, not at all. I feel that everybody . . . being educated myself I say I feel that everybody should know their status and be educated. And take a personal responsibility."

Q6. What are your perceptions of the threats African-American families face when someone they know is infected with HIV/AIDS?

> "Retribution some uh, people with distorted views of what HIV/AIDS is not understanding it being ostracized . . . personal attacks. Uh, people . . . uh—back when it first came up people would actually get violent . . . because you sittin next to me, I might catch it . . . And now that you didn't tell me, you know, I'm gonna wanna, you know, do some, you know, bodily harm."

Q8. What are your perceptions about why African-Americans remain disproportionately affected by the disease HIV/AIDS? "Education (lack of), responsibility, risky behaviors, it can't happen to me, uh, just not paying attention to the world around you."

Participant Xi. Participant Xi stated the following (numbers indicate the interview questions):

Q1. How do conspiracy beliefs of genocide impact your sexual health choices? "Conspiracy of genocide affects my sexual health choices greatly because I am aware of the risk involved . . . causes of different diseases are, you know, from my perspective are man-made and put here as a genocide weapon, something to um, destroy us in a behavior that is a natural behavior, which is sex." (Participant Xi)

Q2. What are your perceptions regarding whether there was a conscious intent to create a virus capable of controlling certain populations, mainly minorities, and if so, how and why?

> "Absolutely so, absolutely so . . . the biggest population in the world are the Asians and the Indians . . . . But I don't see them as a threat to humanity or the world. I pretty much see African-Americans [participant is projecting his own views here] . . . I believe it [virus] was definitely something that was introduced into the community to control the population of um, many of the minorities." (Participant Xi)

Q4. Do conspiracy beliefs of genocide impact your decision to bypass getting screened for HIV/AIDS in order to know your HIV status, and if so, how?

> "Absolutely not! That's probably the one thing that, you know, they really want us to be afraid . . . . So I just feel like we all need to screen ourselves and be aware and, and get rid of the fear." (Participant Xi)

Q6. What are your perceptions of the threats African-American families face when someone they know is infected with HIV/AIDS?

> "The threat of affording the medical bills to sustain um this person. The fear of the unknown, knowing that they're powerless. There's not enough education to where you feel like you can approach HIV and AIDS and feel powerful or in control. It's a powerful disease um, because the education is not there . . . the myth that HIV is a death sentence." (Participant Xi)

Q8. What are your perceptions about why African-Americans remain disproportionately affected by the disease HIV/AIDS?

"Because it was forcefully put in our community . . . this disease has been placed in our community specifically to control the population of the African-Americans and destroy us, as a whole [based on participant's own research of the topic HIV/ AIDS]." (Participant Xi)

Participant Omicron. Participant Omicron stated the following (numbers indicate the interview questions):

Q1. How do conspiracy beliefs of genocide impact your sexual health choices? "They don't affect my sexual health choices very much."

Q2. What are your perceptions regarding whether there was a conscious intent to create a virus capable of controlling certain populations, mainly minorities, and if so, how and why? "There may have been . . . within the past couple hundred years I would have to say naw uh I don't believe that people made conscious intent of creating a virus."

Q4. Do conspiracy beliefs of genocide impact your decision to bypass getting screened for HIV/AIDS in order to know your HIV status, and if so, how? "I don't think conspiracy theory affect my decision to bypass getting screened, I would still get screened anyway . . . I would still get screened anyway. Like I said I work in the medical field."

Q6. What are your perceptions of the threats African-American families face when someone they know is infected with HIV/AIDS?

"You feel out casted, there are families like that where they don't know much about the virus or don't have a lot of background on medical terms or don't understand [lacking education about HIV/AIDS] . . . feel afraid [fear is a threat] . . . if one person has it in the family maybe someone else has it [misconception]." (Participant Omicron)

Q8. What are your perceptions about why African-Americans remain disproportionately affected by the disease HIV/AIDS? "Lack of education, lack of resources, and you know and I don't know what people's sexual choices are like."

Participant Rho. Participant Rho stated the following (numbers indicate the interview questions):

Q1. How do conspiracy beliefs of genocide impact your sexual health choices? "Understanding behavior of uh, uh venereal disease as it relates to sexually transmitted diseases. Um uh causes you to understand what practices or unsafe practices can do to you. Whether they are sexually transmitted base, they can end up being terminal." (Participant Rho)

Q2. What are your perceptions regarding whether there was a conscious intent to create a virus capable of controlling certain populations, mainly minorities, and if so, how and why?

"If you look at your Tuskegee uh, uh, studies and how blacks were infected with uh syphilis. Even if you go back to the uh Mid-Atlantic transfer, uh and you talk about how uh the names of people were taken from them and given new names . . . I would say that it is highly possible that uh a control drug or a drug could be created to say hey let's use this to, to, to control a population." (Participant Rho)

Q4. Do conspiracy beliefs of genocide impact your decision to bypass getting screened for HIV/AIDS in order to know your HIV status, and if so, how? "My beliefs on genocide does not impact because I believe that uh we should be tested, we should be. I am an advocate because I am a social worker, a clinical social worker at that."

Q6. What are your perceptions of the threats African-American families face when someone they know is infected with HIV/AIDS?

"One of the major effects is the participant identifiers avoided using the first letter of all African-American Greek organizations to prevent inadvertent assignment of participants with an identifier of an established and incorporated Greek fraternity or sorority organization of which they might be members., is the unknown of the uh their physical health or the side effects . . . how it impact them economically, how it would impact them socially, how would it impact them in terms of their employment, how would it impact them even from a prestigious perspective . . . as a church member the reaction may be. Example, you may be asked to not participate in a committee that you belong to for a period of time. You may be asked to uh relinquish your role on the choir . . . . People may not come to you and greet you the way that they used to . . . people may feel uncomfortable eating lunch with you, they may feel uncomfortable just talking to you so now the social dynamics change. Even the communications now changes." (Participant Rho)

Q8. What are your perceptions about why African-Americans remain disproportionately affected by the disease HIV/AIDS?

"We don't trust that condoms work . . . we have the idea that um, um it's not gonna happen to me . . . if I'm goin out, umma go out this way [the complacency attitude]. I think a large portion of it is . . . is . . . is . . . is culture, uh, aaaand the . . . the habits of the culture and subculture in our community, in terms of our practices of safe sex or say sexual engagement." (Participant Rho)

Participant Chi. Participant Chi stated the following (numbers indicate the interview questions):

Q1. How do conspiracy beliefs of genocide impact your sexual health choices? "I came up with the stereotype of . . . that I couldn't get it because I wasn't um, homosexual or whatever . . . it made me think whereas I mean protecting myself—doing the right things as far as using a condom and not having multiple sexual partners." (Participant Chi)

Q2. What are your perceptions regarding whether there was a conscious intent to create a virus capable of controlling certain populations, mainly minorities, and if so, how and why?

> "Looking at how AIDS is affecting Africa, in particular, and the people that it is affecting, I mean the thought did come into mind that maybe this disease was created to wipe out the uh African population or even control the growth of the population . . . I do believe that there was an intent for this disease . . . and who it's affecting now? It is affecting the black community and African-Americans in particular."

Q4. Do conspiracy beliefs of genocide impact your decision to bypass getting screened for HIV/AIDS in order to know your HIV status, and if so, how? "No it doesn't affect my decision and one, with me and my current occupation I have to get tested yearly anyway."

Q6. What are your perceptions of the threats African-American families face when someone they know is infected with HIV/AIDS?

> "African-Americans can't afford the care it takes . . . it seems like if I know somebody that has HIV or AIDS in the black community like . . . that's it! It means death . . . if that person is the one that's bring in a lot of the financial or taking care of the financial burden for the family. I mean that could be pretty much it with his family. Yes, devastation." (Participant Chi)

Q8. What are your perceptions about why African-Americans remain disproportionately affected by the disease HIV/AIDS?

> "Lack of education . . . visual media and sex is pushed out to our community like it's just no repercussions behind it I mean . . . not the diseases that are associated with just this free-having sex [more emphasis on pregnancy

and less on STDs, namely HIV/AIDS] . . .
the commercials you see for condom commercials I mean you rarely see 'em. Or, the
AIDS commercials get tested for it . . . you
rarely see it. I rarely see it . . . it's rarely anything about to protect yourself or practicing
to protect yourself [public service commer-
cials]." (Participant Chi)

Summary of male perspectives. Five of seven male
participants (71%) perceived that conspiracy beliefs of
genocide had some influence on their sexual health choices. Of those, two participants (4%) felt that men should
be more vigilant about protecting themselves. One of the
seven participants (14%) felt that young men need to be
taught early the causes and influences of sexually transmitted diseases (STD). Another male participant (14%)
perceived that conspiracy beliefs of genocide greatly influenced his sexual health choices by heightening his awareness of risks, and he strongly believed some viruses were
introduced to the African-American community to cause
harm. Two of seven male participants (28%) did not feel
that conspiracy beliefs of genocide influenced their sexual
health choices at all.

Five of seven (71%) male respondents perceived that a
conscious intent to create a virus capable of controlling certain populations existed, mainly for minorities. One male
participant (14%) reported he had no way of validating

his perceptions on the subject. Another 14% of male participants perceived the action may have transpired to keep minorities out of power. Another male participant stated there have been concerted, conscious actions coordinated to control populations.

A fourth male participant (14%) perceived definitely that a virus was introduced into the black community to control populations, "many of the minorities" (Participant Xi). A fifth male participant (14%) speculated that the spread of HIV/AIDS in Africa mirrors how it is affecting the black community in America, and that this is evidence that some conspiracy has transpired. One of the two male participants who did not believe there was conscious intent to create a virus stated the action did not occur in the case of HIV/AIDS, but did occur in the case of smallpox against the American Indians. The other male participant reported that perhaps this action of some conspiracy to create a virus occurred 100 to 200 years ago.

Some issues emerging from the men's responses echoed some of the women's comments. One of the male participants cited a need for education on the topic of screening. Only one of seven male participants (14%) did not screen for HIV/AIDS, but six of seven (86%) screened for HIV/AIDS. Two of the seven male participants (29%) alleged conspiracy beliefs do not influence their decision to bypass getting screened for HIV/AIDS. One male participant felt that part of the *conspiracy* was allowing conspiracy perceptions to influence their decision.

Some threats faced by African-American families as perceived by the male participants included vicious attacks, bodily harm, ostracism, becoming an outcast, sometimes in their own family, harboring distorted views about HIV/AIDS, retribution, violence perpetrated against the infected victims of HIV/AIDS, not being able to afford medical bills, fear of the unknown, being rendered powerless, lacking in the education of the disease, the myth that the disease HIV/AIDS is a death sentence, and the belief that if one person in the family has HIV/AIDS, perhaps others do.

Finally, two male participants (28%) noted lack of financial support as a threat. Another 28% perceived that African-American families not wanting others to know what they do personally were a reason for the disproportion. Another male participant (14%) alleged African-Americans sometimes do not get questions answered as a factor in the disproportion in numbers of African-Americans contracting HIV/AIDS.

One of seven male participants (14%) said that black men's attitudes toward wearing condoms were a factor. Two out of seven male participants (28%) mentioned lack of education. One male participant mentioned risky behaviors, lack of attention, poor sexual choices, and habits of a culture or subculture as contributing factors. Two of seven male participants (28%) perceived lack of education as a factor in the disproportionate statistics of HIV/AIDS in the African-American community.

Summary

Based on results for the 11 interview questions, six of nine female participants (67%) reported conspiracy beliefs of genocide did not influence their sexual health behavior, and three of nine female participants (33%) reported conspiracy beliefs of genocide influenced their sexual health behavior. Among men, five of seven (71%) reported conspiracy beliefs influenced their sexual health behavior, and two of seven (29%) men reported conspiracy beliefs of genocide had no influence over their sexual health behavior. These results indicate that 10 of 16 (63%) participants report having perceptions concerning conspiracy beliefs of genocide have some influence on their sexual health behavior, and male participants were the majority with this view.

Only two of the 16 participants, both women, mentioned trust in medical professionals as an issue, and among the two, each was split on whether to trust or not trust doctors or the medical community as a significant factor when seeking medical attention. The two women's answers about trust in the medical community are conclusive for the study findings. Among female participants, no majority consensus emerged regarding lack of trust in the medical community or medical doctors. One of nine female participants expressed strong distrust in medical doctors. The majority of female participants did not believe a conspiracy existed, or that such conspiracy affected their sexual health choices.

The data further suggest that the HIV/AIDS virus affects African-Americans disproportionately because African-Americans fear the results of the HIV/AIDS test and so do not seek screenings because they do not want to know their status. Three of nine female participants (33%) stated that African-American men do not like to wear condoms, and two of the nine female participants (22%) believed African-Americans feel they are immune to catching the virus HIV/AIDS or that it could not happen to them. One female participant stated lack of knowledge *early enough* to not infect an exponential number of other people contributed as a significant factor for the disproportion. Christian values seem to conflict with reality, suggesting that some men who remain faithful to their wives but succumb to the pressure of pleasure are not prepared to protect themselves because they may have had every intention to remain faithful.

Among participating men, lack of trust in the medical community or medical doctors did not emerge as an important factor. Two of seven men (29%) harbored distrust of doctors in general. Two of seven male participants worked in the medical community.

The majority of male participants, five of seven (71%), did not believe a conspiracy existed or that such conspiracy affected their sexual health choices. One male participant cited the dire need for men to protect themselves and the person they love. Another male participant believed a conspiracy existed and that lack of education was a barrier to counteracting the influence of this conspiracy.

One male participant stated that viruses were man-made, were introduced into the black community to cause harm, and were used as weapons. For Question #8, the collective consensus among men included lack of financial support, privacy concerns, attitudes about condom use, lack of education, engaging in risky behaviors including multiple sex partners, and refusing to screen in order to know one's HIV status as drivers to the disproportion of HIV/AIDS in the African-American community. None of the nine women and seven men in the study mentioned vaccinations or male circumcision as methods of prevention for the HIV/AIDS virus.

5

IMPLICATIONS AND
RECOMMENDATIONS

THIS QUALITATIVE, HERMENEUTIC, phenomenological research study generated several findings to be added to an existing body of research on HIV/AIDS. The study data offer unique perspectives with serious implications for leaders in health care treating HIV/AIDS in the African-American community. The goal of this research study was to probe the stream of consciousness of African-Americans to address the following two research questions:

R1.   What are the lived experiences and perceptions of African-Americans pertaining to conspiracy beliefs and the influence of such beliefs on their sexual behaviors?

R2.   What influence, if any, do conspiracy beliefs held by African-Americans have on their decisions associated with seeking medical attention?

The data provided the perspectives of African-American men and women on HIV/AIDS.

Analysis and Discussion

IDENTIFICATION OF MAJOR THEMES

The literature review suggested stigma and shame were prevalent, as Tomaszweski's article in the NASW (2012) described how the stigma and shame of HIV/AIDS among minorities and the poor create social obstacles to health surveillance and prevention tactics. Sixteen African-Americans, recruited from across the country through mailings and referrals, participated in the study. The results yielded 17 themes, reduced to the following five overarching themes: (a) belief that a virus was created; (b) homosexual behavior as an implication; (c) lack of access to health care as a threat; (d) lack of education on HIV/AIDS; and (e) stigma, ostracism, shame, shunning, and stereotyping of persons diagnosed with HIV. Perceptions about a conspiracy to control the African-American population seemed to have no influence overall on sexual health behavior, including whether or not to seek medical attention. According to the participants, lack of health care derives from lack of eligibility to receive treatment based on preexisting conditions such as HIV/AIDS, as well as shame and stigma that encourage denial and poor decisions about whether to use condoms or seek screening for HIV status. Based on the participants' responses, lack of education about HIV/AIDS is a major factor in the higher-than-typical, sustained epidemiology of HIV/AIDS among African-Americans.

From the literature review, the Tuskegee experiments appeared to have laid a foundation for conceptualizing HIV/AIDS as a virus that American scientists might have created (Thomas & Quinn, 1991). Other feedback from participants suggests that the smallpox incident perpetrated against American Indians by white Americans continues to support belief in conspiracy. In this study, lack of trust in the medical community emerged as a factor in a decrease in health screenings, which results in fewer adherences to retrovirus medication, but trust or lack of trust did not emerge as a trend (Saha, Jacobs, Moore, & Beach, 2010).

Participant Beta voiced a strong positive opinion on trust in medical professionals. This result might suggest that lack of trust in medical personnel, such as doctors and nurses, may be dissipating. More research about attitudes and trust in doctors and nurses treating African-Americans with infectious diseases may be warranted.

LaForce (2006) provided information about the MSM culture and explained psychological inhibitions leading to risky behaviors. Alcohol and drugs are a major factor in engaging in unsafe sexual actions. Defenses that prevent individuals from engaging in unsafe sex acts seem to be inhibited by the effects of alcohol and illicit substances. In the current study, the cost of care and lack of health care were strong themes.

Social pressures may be suggesting that African-Americans must be violating sexual identification and heterosexual gender norms already set in stone by society and that a diagnosis of HIV/AIDS automatically confirms sexual identity. According to Kallings (2008), drug and alcohol abuse drives perceptions about character among individuals with an HIV/AIDS diagnosis, and ostracism results from the perceptions of practices that prohibit African-Americans from wanting to screen or openly have a conversation about the subject of HIV/AIDS.

Overcoming such perceptions would likely lead to more African-Americans being open to screenings, seeking care, and protecting themselves and others. More faith-based organizations must be willing to address HIV/AIDS as a public health crisis in the African-American community, and families must support members who become diagnosed. As a disease, HIV/AIDS must be considered treatable if caught early, not an automatic death sentence.

Theme 1. Repeated theme patterns emerging from the data included HIV/AIDS as a virus that American scientists created. Participant Beta stated, "I believe there was conscious intent to create a virus capable of controlling blacks." Participant Eta expressed, "A virus may have been created to control a race." According to Participant Pi, "Human experiments have been taking place that allow a germ to spread and fester . . . . White settlers were guilty of such atrocities." Participant Chi stated, "There may have been an attempt to wipe out the African population through AIDS." Participant Rho stated, "History tells you Tuskegee experiments took place to possibly control a population." Participant Lambda shared, "I am one who believes viruses were created to control population, and they grew out of control." According to Participant Xi, "HIV/AIDS was introduced to the African-American communities to control the population." Based on this pattern of responses, 44% of the participants concurred with the perception of HIV/AIDS is a virus that may have been created.

Theme 2. Theme 2 included concepts on homosexual behavior, especially with men having sex with men (MSM). Participant Beta explained, "Down-low brothers (men who have sex with men, or MSM) become infected and pass HIV/AIDS on." Participant Epsilon stated, "Fear of being seen as engaging with the 'down-low' culture." Participant Chi believed, "AIDS in the mid-eighties affected the homosexual population initially." Participant Tau said, "To have HIV/AIDS implies one is gay." Thirty-one percent of respondents had concerns over the perception of being seen as homosexual or *gay* and associating such perceptions with HIV. None of the nine female and seven male participants mentioned vaccinations or male circumcision as methods of prevention for the HIV/AIDS virus. This may represent a gap in research, and further

investigation on attitudes about circumcision and HIV/
AIDS vaccination is warranted.

Talking about a disease that carries great stigma, shame,
and other social, financial, legal, and personal challenges
is difficult. Among social issues arising from this research
was the lack of ways to break barriers to communication
with a lover or spouse. To better understand this barrier
in context, the researcher viewed the documentary End
Game: AIDS in Black America (Simone, 2012). End
Game provides a riveting account of the social issues black
people face with regard to HIV. A patient in the film vis-
its the doctor for persistent flu-like symptoms when her
male partner insists. A positive diagnosis for HIV results
after no evidence is found to support other diagnoses. The
woman's male partner in the film keeps his HIV status
secret.

Theme 3. Theme 3 included issues of health care,
access to health insurance, and cost of care. Participant
Eta stated, "Lack access to health care." Participant Pi
shared, "African-Americans' lack of access to health care."
Participant Rho believed, "Not having the funds to support the care needed is a threat to African-American
families." Participant Chi suggested, "Affording care for
the black community with relatives afflicted with HIV/
AIDS." According to Participant Lambda, "Lack of health
care in African-American communities drives disparity of
HIV/AIDS . . . . African-Americans in rural areas where
income is lower health care is scarce." Participant Nu

stated, "Lack of African-American access to health care." Participant Chi said, "Death due to bad health from believing in conspiracy theories and allowing fear to impact health decisions." Participant Omicron expressed, "Lack of health insurance, and sometimes the statistics don't reveal the full extent to which we are affected." The results for Theme 3 indicate that 50% of the respondents consider issues related to lack of health care, insurance, and cost of care as barriers.

Theme 4. Theme 4 is a pattern of responses about lacking education. Participant Beta stated, "Education is lacking . . . . Uneducated about transmission." Participant Gamma expressed, "Lack of access to appropriate, affective health education." According to Participant Chi, "Education (realistic) i.e., epidemiology on who it's affecting most. Education and awareness on the epidemiology of society most affected. African-Americans' lack of education and awareness." Participant Iota said, "African-Americans need more education on epidemiology on HIV/AIDS . . . African-Americans need more education on how HIV is transmitted." Participant Lambda stated, "Education levels down." Participant Mu believed, "Education on personal health." Participant Nu stated, "Better education will lead to better sex health choices. Education as defense." Participant Xi explained, "Lack of education, control and power . . . African-Americans need more education about HIV/AIDS." Sixty-nine percent of the participants cited education or lack of specific

education on HIV/AIDS in the African-American community as a major problem. Education on the epidemiology of HIV emerged as a topic that may have lasting
impressions and lead to change.

Theme 5. Theme 5 is stigma, ostracism or shunning, and shame or stereotyping. Participant Gamma
stated, "If there is someone in the family who becomes
infected with HIV and AIDS, that person may want to
disassociate from the family, or the family may want to
shun that person on their own." Participant Pi expressed,
"There would be shame and guilt to face." According to
Participant Iota, "African-American families are/feel ostracized." Participant Mu stated, "Being ostracized and
looked upon as something dirty because the disease un-
fortunately in the African-American community is still
not understood." Participant Nu believed that "distorted
views of what HIV/AIDS is, not understanding it, being ostracized, things like that. If I was to say, threats."
Participant Omicron stated, "Shun the family socially."
Ostracism and shunning presented as a major theme.

Fifty percent of participants in this study reporting that
stigma, shame, ostracism, shunning, and stereotyping created
fears about HIV/AIDS, and these fears prevent people from
screening or taking any action for treatment of HIV that
might identify them as being infected. This high percentage
was critical to understanding the current epidemiology and
the historically escalating, though leveling, incidence of HIV/
AIDS in the African-American community.

DISCUSSION OF MAJOR THEMES

Each of these themes is discussed in detail below.

HIV/AIDS is a disease perceived as having been created (Theme 1). Some participants (44%) perceived that HIV/AIDS is a virus that may have been created and evolved in a pandemic. The origins of the disease HIV/AIDS remain unknown. Specific evidence on the spread of HIV/AIDS points to increased travel across the continents as well as practices in Africa that exposed humans to blood contaminants in animals they butchered for bush meat (Wolfe, 2011). The Tuskegee experiments (Thomas & Quinn, 1991) and HeLa Cells experiments (Lucey et al, 2009) are well-documented scientific activities among American scientists that provide historical background for minorities.

Being perceived as homosexual (Theme 2). Data from the CDC (2007) indicate higher epidemiology in the MSM culture than in other groups. In this research study, 31% of participants expressed their perspectives on homosexuality. J. L. King (2005) found that men who have sex with other men remain relatively isolated from the general public, which presents issues for prevention, outreach efforts, and surveillance. Men in the MSM culture do not necessarily identify as being homosexual. The profile of MSM is straight men who engage in sexual activity with other straight men. Activities such as screening, obtaining education, or identifying with HIV/AIDS have implications about homosexual orientation for MSM. "Nearly a quarter of black HIV positive men who have sex with men (MSM) consider themselves heterosexual" (CDC, 2007, pg. 1).

Lack of access to health care (Theme 3). One of the discriminatory markers for health-care disparities before the passing of the Affordable Care Act or ObamaCare in 2008 (American Public Health Association, 2012) upheld by the United States Supreme Court in 2010, included preexisting conditions as a reason for refusing treatment. The new legislation is critical for the issue of health disparities in terms of preventing the spread of HIV/AIDS and other STIs among African-Americans, the poor, and individuals without health care. Among the participants in the present research study, 50% agreed that lack of access to health care is a detriment and a significant threat to African-Americans who seek to prevent the spread of HIV/AIDS. This information explains, in part, the higher than typical epidemiology of HIV/AIDS among blacks.

Lack of education on HIV/AIDS (Theme 4). Not having enough education on HIV/AIDS was of particular concern for 69% of participants queried in this research study. According to the CDC (2010), health disparities based on racism, poverty, lack of success in accessing health care, and lack of education about HIV/AIDS and STDs are a barrier to care. Gender differences factor in the incidence of HIV/AIDS. The MSM population reflects consistently higher epidemiology than women (CDC, 2010).

HIV and stigma/shame (Theme 5). The HIV/AIDS disease affects minorities and poor people more frequently than individuals in other groups. Stigma and shame are barriers for appropriate levels of care. Stigmatization about HIV/AIDS occurs through legislation, disclosure issues, and quality of life (Tomaszweski, 2012). Current researchers identified correlations among the five overarching themes of this study.

According to the participants' responses, many African-Americans experience stigma and shame (50%) with regard to HIV/AIDS. Ronald Johnson from CNN (2012) led a conversation on the incident of the Stonewall riots, in which homosexuals were targeted through stigmatization, victimization, and extreme violence in 1969. Recent legislation from the NAACP and President Barack Obama on same-sex marriage, developments on the Defense of Marriage Act, California's Proposition 8, and federal court rulings in favor of same-sex marriage provide a snapshot of progression in America toward tolerance and equality.

Summary of Major Themes

The table below shows the five major themes that were extrapolated from the data.

Table 1: Participants' Responses to Five Major Overarching Themes from Interviews

Participant[a]	Belief a virus was made	Homosexuality	Lack or access to health care as a threat	Lack of Education	Stigma, Ostracism, Shame, Shun, Stereotyping
Beta	Yes	Yes	No	Yes	No
Gamma	No	No	No	Yes	Yes
Epsilon	No	Yes	No	No	No
Theta	Not Sure	No	No	No	No
Eta	Not Sure	No	Yes	No	No
Iota	Yes	Yes	No	Yes	No
Lambda	Yes	No	Yes	Yes	No
Mu	No	No	No	Yes	Yes
Nu	No	No	Yes	Yes	Yes
Xi	Yes	No	Yes	Yes	No
Omicron	No	No	Yes	Yes	Yes
Pi	No	No	Yes	Yes	Yes
Rho	Yes	No	Yes	No	Yes
Tau	Yes	Yes	Yes	No	Yes
Upsilon	No	Yes	No	Yes	No
Chi	Yes	No	No	Yes	Yes
Total	Y= 7	Y=5	Y=8	Y=11	Y=8
	N= 7	N=12	N=8	N=5	N=8
	44%	31%	50%	69%	50%

[a] To protect participant confidentiality, pseudonyms are used to replace all names.

Table 1 includes a breakdown by percentage of the five overarching themes that emerged from participants' responses to the 11 interview questions. Lack of education, with 69%, was the most important concern for all respondents. Stigma, shame, and being perceived as homosexual (50% each) were other strong themes emerging from the study.

Recommendations

The disease known as HIV/AIDS plagued the American population for decades before African-Americans became disproportionately affected (Wolfe, 2011). In the early 1980s, the white male homosexual community was severely affected and was almost exclusively depicted through CDC (2008) epidemiology studies. Members of the homosexual community lobbied and fought for funding and care through social movements (ACT UP, 1998). According to the CDC, from 1985–2005 the epidemiology of HIV/AIDS for white male homosexuals drastically declined, and the percentage for African-Americans soared.

The incubation period for HIV/AIDS can span from six months to one year (Njororai, Bates, & Njororai, 2012). Despite abundant evidence of efforts to contain the virus, epidemiology has remained high among African-Americans. Social movements and faith-based efforts support existing interventions, including lobbying. Njororai et al (2012) described Voluntary Action Centers (VAC) as confidential means to battle HIV/AIDS. The VACs could be funded and strategically situated all across America, particularly in states or cities with the highest rates of HIV/AIDS.

RECOMMENDATION 1

The first recommendation addresses Theme 4, lack of education and attention for HIV/AIDS. More attention needs to be directed to existing social movements, with a focus on actively working with leaders to identify barriers to prevention. Of primary interest in this goal is the number of newly infected African-Americans across the nation and in highly populated areas, where the epidemiology for HIV/AIDS remains prevalent.

The organization Act Against AIDS (2012) is reporting a 91% prenatal transmission rate. "One Test Two Lives" is a program specifically targeting pregnant women, providing educational material and voluntary screening. A recommendation is to institute a campaign to launch a "One Test Two Lives" program in every state. Every pregnant woman should be afforded the opportunity to learn of her HIV status. Funding and support should remain constant for these efforts. According to CDC (2008), "Surveillance estimates indicate that 96-186 infants are infected with HIV per year" (pg. 1). These statistics expose the need to broaden opportunities for pregnant women to learn of their HIV status and of the status of their fetus, especially in rural areas and states with higher incidence of HIV/AIDS.

Each state should create an HIV/AIDS handbook listing resources for clinical treatment and screening, as well as an explanation of emergency procedures, including the people to notify and how to notify them. Education is a strong theme in the current study. Education focusing on the epidemiology of HIV/AIDS in the African-American community must be available and extend to access to care, or incorporate education on the epidemiology of HIV/ AIDS in the African-American community with interventions for care. The goal is to increase medical care for newly diagnosed individuals, prevent new cases from emerging, and decrease the possibility of an infected person transmitting the virus. To achieve this, clinicians and lawmakers must broaden confidential screening, education, and focus on epidemiologic numbers.

Education on HIV/AIDS should target adolescent children through schooling, inclusive of high school. Such education should be mandatory. The information provided should include (a) history of HIV/AIDS; (b) transmission; (c) the etiology of HIV/AIDS; (d) factual, historical information about the Tuskegee studies; (e) epidemiology among the African-American community as related to HIV/AIDS; (f) clinical trials and treatment; and (g) protection methods using layers of protection, not solely relying on condoms, to afford maximum protection against the virus.

Education on medication efficacy with data generated from clinical trials and pharmaceutical continuing medical education (CME) lectures for teachers and parents should be included in appropriate curricula. Ramifications of unprotected sexual encounters and details about risky behaviors need to be emphasized and provided at varying educational levels, where appropriate. An aspect of the educational curriculum should be dedicated to track the sequence of events from the time of infection to the time of effective management and control of HIV/AIDS.

Action plans should offer the PBS documentary End Game: AIDS in Black America (Simone, 2012) as required viewing in high schools and colleges and expand HIV screening sites across the nation to a 10- to 15-mile radius. Prevention efforts can minimize pandemics like HIV. Although Hawaii's HIV/AIDS rate is 1%, maintaining a low rate of new HIV cases will require interventions, such as screening availability in rural areas. At the time of the study, the nearest testing site to some residents in Waianae, Hawaii, is a 40-mile journey. Action plans should include offering screening sites for all rural areas, where prevention is as important as in densely populated areas, where incidences of HIV are higher.

Another aspect of HIV/AIDS prevention highlighted in the literature is poverty as a factor in higher incidence of the disease. Williams (2003) noted, "Despite the growing number of this ethnic minority in the US population, the growth in the number of minority health care professionals is not keeping pace" (pg. 8). African-Americans may be more invested in seeking care by professionals of the same ethnic background, which would lead to reducing the incidence of the disease.

Low education, poverty, myths, and other barriers prevent progress in reducing disparities in the incidence of HIV/AIDS. Williams (2003) suggested "parental involvement in all health, education, social and cultural HIV/AIDS intervention programs" (pg. 289). Based on the participants' responses in this research endeavor, lack of education was a major theme, along with shame and ostracism. The Office of National AIDS Policy (ONAP) failed to include education as part of its three-pronged strategic approach.

President Barack Obama highlighted goals for ONAP (2012) as (a) reducing new HIV infection rates; (b) improving health care and treatment efforts; and (c) reducing health-care disparities, including preventing insurance companies from denying treatment and care for preexisting conditions, namely HIV/AIDS. Unless federal and state governments are mandated to include education as a means of reducing new cases, gaps will continue to exist. The ONAP agenda does not include providing HIV/AIDS education in public school curricula (see Appendix G).

RECOMMENDATION 2

In referencing Theme 3, the second recommendation targets health care and health-care disparities. To ensure that HIV/AIDS is not being spread from patient to patient, dentists should mandate patients to be tested for HIV. Disparities of HIV/AIDS incidence among African-Americans have remained disproportionate since 1992 (CDC, 2008). Crucial feedback in the present research resulted in recommendations to assist current leaders with efforts to bridge the disparity and decrease the incidence of HIV/AIDS in the African-American community. The new health-care law went into effect in 2014 and will include language to target activities that have discriminated historically against the poor, and people with preexisting conditions such as HIV/AIDS.

The theme of health care, or lack of health care, emerged from the current study data. Care for HIV/AIDS is further compromised by health disparities among African-Americans. Several participants in this research project mentioned lack of health care and affordability of care for HIV/AIDS. The recommendation is to mitigate health-care disparities by targeting areas with a high incidence of HIV/AIDS in 2014, when the Patient Protection and Affordable Care Act (ACA) was implemented (Institute of Medicine, 2012). Besides providing health care for those who cannot afford to purchase insurance, developing awareness by expanding public service campaigns for STD protection is necessary.

Leaders of states with a high incidence of HIV/AIDS should launch similar campaigns to protect against STDs, specifically HIV/AIDS. For rural and densely populated areas with a high incidence of HIV rates, consistent attention to linkage to medical care, including access to medication, is a logistical problem. Keeping infected individuals healthy requires easy access to care and medication. Strategic plans to enable access would minimize fear for some who are reluctant to know their status because they are afraid they would not be able to care for themselves if they had contracted the disease.

According to the Honolulu Star Advertiser (2012), "The U.S. health care system squanders $750 billion a year—roughly 30 cents of every medical dollar—through unneeded care, byzantine paperwork, fraud, and other waste" (pg. 3). This insight provides a clear direction to address waste in the American health-care system and to use funds more efficiently to combat HIV/AIDS in the African-American community and in America. Health organizations should campaign for early intervention upon an HIV-positive status to prevent further spreading. Since support and funding for Planned Parenthood clinics include counseling on HIV/AIDS, the programs should remain a priority in the fight to decrease the incidence of HIV/AIDS.

RECOMMENDATION 3

The third recommendation stems directly from Theme 5, which addresses stigma, shame, ostracism, and shunning as barriers to prevention of HIV through education and more research. Removing misnomers and false beliefs can be a challenge, but through constant and relentless efforts to educate the public about various aspects of the disease HIV/AIDS, the barrier of stigma, shame, ostracism, and shunning could be reduced. According to Buseh (2006, pg. 3), "HIV stigma is widely regarded as a major obstacle to effective HIV prevention, risk reduction, testing, and treatment. Research is urgently needed to anticipate, understand, and combat stigma in the African-American cultural context." More research on cognitions of stigma, shame, ostracism, and shunning is needed to further understand how to address these issues.

According to Greenberg et al (2009), the epidemic of HIV/AIDS among African- Americans in the nation's capital is comparable to some sub-Saharan countries in Africa, and some large U.S. cities such as Chicago and Philadelphia. Health-care leaders' response to the AIDS crisis has been to forge a partnership between the Washington, D.C. Department of Health's HIV/AIDS Administration and the George Washington University School of Public Health and Health Services. Surveillance activities were funded and improved, and the partnership was so successful that it was broadcast in the local media.

Another movement born and nurtured in Washington D.C. included a screening program called "Come to-gether DC—Get screened for HIV" launched in 2006. Washington D.C., the nation's capital, has the highest epidemiological rates of HIV at 100,000 per 117.7 population, second only to Maryland. The high rates of HIV for Washington D.C. and other cities such as Maryland, Philadelphia, Chicago, Detroit, Louisiana, Atlanta, and New York have remained unchanged over the course of approximately a decade. The "Come together DC—get screened for HIV" campaign is successfully raising awareness to the high rates of HIV among African-Americans and is a flagship movement emphasizing the importance of screening for HIV. Transmitting HIV can only occur if the virus is present. Therefore, it is crucial to know your HIV status and the HIV status of your partner. Furthermore, the key to preserving health when the infection enters the body is to diagnose HIV at its earliest stage to prevent further immune compromising.

In 2003, President George W. Bush's administration unveiled policy with the goal to significantly impact AIDS through the President's Emergency Plan for AIDS Relief (PEPFAR), contributing over 148 billion dollars at home and worldwide. Although this global effort is making a great impact, the epidemiology of HIV/AIDS among black adolescents and men in America remains consistently higher than any other race/ethnicity.

The effort to stem the tide of HIV/AIDS epidemiology in the black community through the screening program included: (a) a behavioral surveillance system that focused on healthy relationships, (b) condom-distribution programs making free condoms available in places such as hair salons and barbershops, (c) targeted services for higher-risk groups and for populations in dense rural areas, (d) harm-reduction programs, and (e) programs targeting youths (2009). Every state needs an HIV/AIDS Advisory Board/Commission to advance its agenda for HIV/AIDS. Since many public schools do not endorse educational modules on HIV/AIDS, private and public counseling and therapeutic personnel may need to spearhead the fight against HIV/AIDS. Social movements like these need priority funding and approval in every state, or in states experiencing high incidence of HIV/AIDS.

Jacobs and Johnson (2007) discussed another strategy, the social movement Treatment Action Campaign (TAC). The group works with local media to highlight the impact of HIV/AIDS in Africa as a way to make it public and transform it into a political issue, forcing leaders to address the public health crisis of HIV/AIDS in Africa. Through these efforts, stigma, shame, ostracism, and shunning could be addressed and removed as barriers to intervening and treating HIV/AIDS in America.

According to Johnson (2012) in a CNN report, AIDS United works with other organizations to provide care for clients and expose some of the ignorance that exists, including being afraid to hug someone afflicted with the HIV/AIDS virus. Johnson (2012) revealed that 1.2 million Americans live with HIV/AIDS. New advances in treatment can extend the life of a 25-year-old by up to 50 years. Demystifying HIV/AIDS through counseling, education, and public service awareness, as a disease found predominantly in gay communities and as one that is a death sentence, would address issues of stigma, shame, and ostracism. According to Johnson (2012), the Federal Drug Administration (FDA) recently approved a drug for prevention of HIV/AIDS. The drug Truvada is a combination of two drugs, Emtriva and Viread, given once a day. Clinical trials have shown effectiveness in preventing HIV/AIDS. This therapy may greatly increase the likelihood that individuals will seek screening and may also assist with curbing incidence of new cases. In addition to Truvada, the 2014 Obama HealthCare initiative's new policy on denial of coverage due to preexisting conditions should greatly assist in reducing epidemiological statistics for African-Americans over time.

RECOMMENDATION 4

Theme 2 pertains to being perceived as engaging in homosexual behavior, especially the behavior of the "down-low" culture. When legal, educational, and religious institutions can be more open about the realities that exist in America in the changing paradigm of sexual relations, the culture as a whole can begin to address health concerns more readily. Educational curricula should include the following books on their reading lists: On the Down Low, by J. L. King; On the Up and Up, by B. Browder and K. Hunter; 20 Warning Signs of Down Low Brothers, by Nubia; and A Time to Embrace: Same-Sex Relationships in Religion, Law, Politics, by William Johnson, and this book; AIDS Crisis Among African-Americans: The CDC reports HIV/AIDS for African-American Men are 70% of New Infections.

Intimacy Social Matrix: ABC Beat the Street Model

The following describes a model by the researcher based on the findings of the study. The HIV/AIDS virus has plagued the African-American community for over two decades, according to the CDC (2008) surveillance studies. Many approaches to decreasing the number of new cases include needle exchange programs, heightened awareness campaigns for prevention, increased funding, counseling, and other initiatives. Some strategies have not been implemented.

The Intimacy Social Matrix model depicts approaches that leaders in states with high incidence of HIV/AIDS should consider. In Target A, state-by-state yearly summits reflect commitment to following epidemiological numbers of newly infected HIV. Individual state health agencies compete for funding by requests for proposals (RFPs) to coordinate summits. In Target B, schools promote prevention and education curriculum starting as early as the Department of Education will allow in each state. In Target C, State Health and Human Services Systems receive adequate funding to (a) coordinate initiatives supporting evidence-based clinical interventions that promote protection from HIV and community education and outreach in environments frequented by vulnerable individuals and (b) provide jobs to assist in finding remedies to socioeconomic disparities and targeting disparaged groups such as homosexual males and prostitutes, and rectify gender inequality, all of which drives the incidence of

HIV in the African-American community.

The intimacy social matrix model depicts stages of appropriate social discourse. The goal of the intimacy social model is to provide a targeted, pragmatic approach to social cues for young men and women who become intimate. The model may be well-suited for teenagers as a learning tool to guide them when making future decisions about intimacy choices. An outcome of this study is to ensure all schools from middle through college are required to teach a curriculum dedicated solely to concepts on HIV/AIDS, with emphasis on prevention, including screening and issues related to sexual health behavior. To show the impact of HIV/AIDS, schools should dedicate learning material on epidemiological trends in areas with a high incidence of HIV/AIDS. Teaching young adults and children more effective ways of communicating socially could ultimately lead to making better choices and charting the course for different outcomes for HIV/AIDS prevalence.

In addition to teaching children and young adults better ways of communicating, strategies such as parental collaboration and chaperoning could alter expectations from society regarding cultural norms for sexual intimacy. The model below provides an alternative that, when practiced, could ease the pressure for effective communication among teens and young adults. Figure 7 is an example of a social matrix illustrating age-appropriate dating behavior that leads to protection from STIs (author's original work).

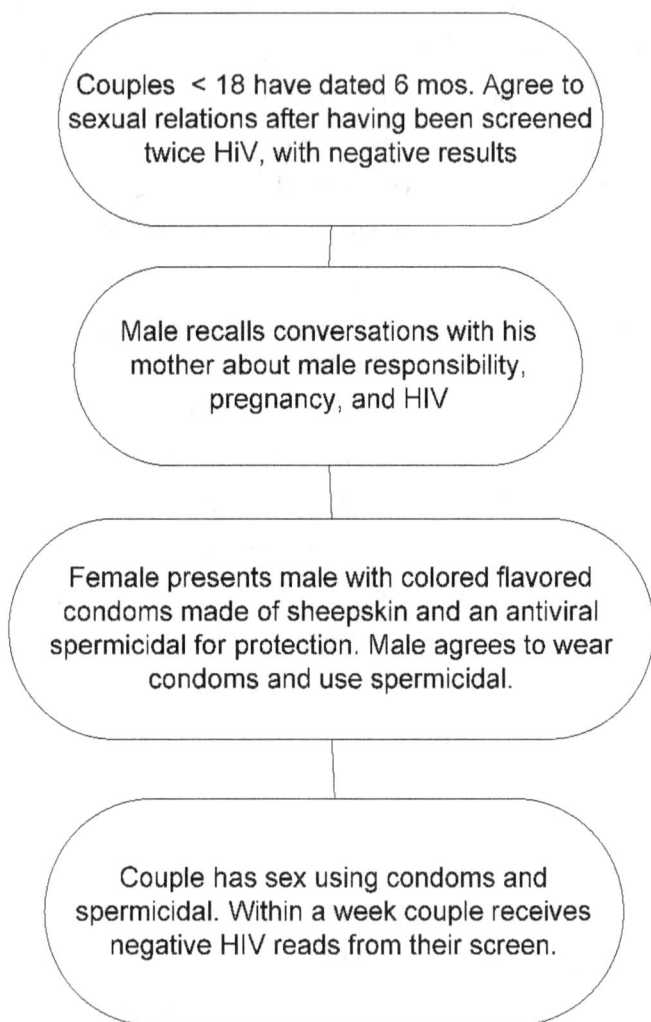

Figure 7. Intimacy social matrix model.
Note: Author's original work.

Recommendations for Future Research

The HIV/AIDS virus in the African-American community is essentially a numbers game at this point in history. The most disturbing feedback stemming from the research interviews came from two female participants with opposite points of view. Participant Beta described personal experiences with medical professionals that led her to a primary set of beliefs. Participant Upsilon, two decades younger, shared thoughts on hearing peer reactions and perceptions on HIV/AIDS related to testing, receiving a diagnosis, not believing the results, and perceiving that the effects of the medication given to control HIV/AIDS hamper cure.

Participant Beta noted:

"My thoughts on conspiracy of genocide and how it affects my sexual health choices are as follows: For one thing, 'genocide' is a word that was originated in 1944. Before 1944 the word did not exist. But for my race of people, African-Americans, there has been a conspiracy ever since the beginning of time, as far as back as I can remember, to limit as many black babies from being born as possible." (Participant Beta)

Participant Upsilon stated,

"Ummmm (hesitation). I have other thoughts . . . . I know there are among, I guess . . . . I don't know if it's just people my own age there is the thought of not getting tested because of more and more African-Americans are coming up with positive HIV tests and being treated with HIV . . . there's a thought, I haven't read the research, just seen where people are writing articles that the HIV positive uh test results are coming up positive when you're negative and that the treatment that they give you to fight HIV causes AIDS, and I know a lot of people who feel that way . . . . Um, it is. I think it definitely gives a mixed message." (Participant Upsilon)

The feedback from participants Beta and Upsilon highlights the need for future research. Further studies are needed on the experiences with medical professionals of African-Americans over age 50 and on attitudes about doctors to gauge how education and training on behalf of the medical community might mitigate trust issues. Future studies could address ongoing measurements of trust in medical providers among women over 50 in areas of the country severely affected by HIV/AIDS.

The feedback from participant Upsilon was the most troubling account of attitudes and perceptions on HIV/ AIDS among African-Americans with regard to medical interventions such as screening and treating HIV. Young people refusing a screening because of their perceptions about the results not being true and that the effects of the medication to treat HIV create HIV is tragic. The two participants had a 20-year age difference. Some research endeavors ought to focus squarely on issues raised during the interviews with the two participants, particularly that HIV/AIDS results reporting as positive are negative and that the medication given causes AIDS. The literature does not provide information about the sequence of events from the time of infection to the time when the disease can be managed.

Summary

Based on responses from the participants in this study and the latest epidemiology data on African-Americans and HIV/AIDS, more financial commitment should originate from Washington D.C. to fight against HIV/AIDS in America, with an emphasis on addressing epidemiological numbers of newly diagnosed cases among blacks. One of the epidemiological goals should be to decrease the number of new cases of HIV being reported among black people in America from 44% to 2%, and later from 2% to 1%. The incidence of HIV among individuals living in regions with high HIV/AIDS incidence should be approximately the same in all states. This goal could be achieved by targeting the most severely affected states and implementing aggressive goals, such as screening pregnant mothers as early as possible.

In recent news, the Huffington Post (2012) published an article about doctors at the University of Mississippi declaring a newborn "functionally cured" of HIV/AIDS. The event may be critical for prevention of HIV/AIDS with a focus on screening and outreach.

Only 19% of Americans living with AIDS receive care (Johnson, 2012). Social changes and more acceptance and tolerance may be a catalyst to decreasing the shadow of shame and stigma that prevents people from seeking care.

Should the recommendations noted contribute optimally to the existing body of knowledge and research, transformation may occur and validate the current findings. The subjective nature of qualitative research makes identifying measures of validity difficult. According to Cho and Trent (2012), "Replicability, testing hypothesis, and objective procedures are not common terms in qualitative researchers' vocabularies" (pg. 319). This view might suggest that the subjective nature of the phenomenological approach eludes attempts to obtain verification.

A link between claims and evidence can validate participants' ontological views. Researchers' claims must correlate with participants' reality, consisting of cultural, social, economic, emotional, or religious constructs. A tight link between findings and the content of participants' narratives ensures high validity. None of the respondents in the study mentioned how frequenting brothels or cultural and religious practices indoctrinating polygamy can help contract or spread HIV/AIDS when precautionary measures are ignored. The impact of these examples on transmission of HIV/AIDS could be daunting. Further research into each of these examples could add valuable social and scientific insight. The International Business Times (March, 2013) published the contributions of ex-Surgeon General C. Everett Koop as the first surgeon general to use his office to rigorously raise awareness of HIV/AIDS while he served.

REFERENCES

American Public Health Association. (2012). *ACA basics and background.* Washington, DC. Retrieved from http://www.apha.org/advocacy/health+reform/ACAbasics

Ary, D., Jacobs, L., & Sorensen, C. K. (2006). *Introduction to research in education* (9th ed.). Belmont, CA: Wadsworth/Cengage.

Bandura, A. (1977). *Social learning theory.* Englewood Cliffs, NJ: Prentice Hall.

Basken, P. (2011, December 11). Flawed 1940s study in Guatemala could mean new research rules for universities. *Chronicle of Higher Education.* Retrieved from http://chronicle.com/

Beck, A. T., Freeman, A., & Davis, D. D. (2004). *Cognitive therapy of personality disorders* (2nd ed.). New York, NY: Guilford Press.

Bing, E. G., Bingham, T., & Millett, G. A. (2008). Research needed to more effectively combat HIV among African-American men who have sex with men. *Journal of the National Medical Association, 100,* 52-56. Retrieved from http://www.nmanet.org/

Blankenship, K. M., Smoyer, B., Bray, S. J., & Mattocks, K. (2005). Black-White disparities in HIV/AIDS: The role of drug policy and the correction system. *Journal of Health Care for the Poor and Underserved, 16*(4), 140-156. doi:10.1353/hpu.2005.0110

Boer, H., & Mashamba, M. T. (2005). Psychosocial correlates of HIV protection motivation among Black adolescents in Venda, South Africa. *AIDS Education and Prevention, 17*, 590-602. doi:10.1521/aeap.2005.17.6.590

Bogart, L. M., & Bird, S. T. (2003). Exploring the relationship between conspiracy beliefs about HIV/AIDS to sexual behaviors and attitudes among African-American adults. *Journal of National Medical Association, 95*, 1057-1068. Retrieved from http://www.nmanet.org/

Bogart, L. M., & Bird, S. T. (2005). Are HIV/AIDS conspiracy beliefs a barrier to HIV prevention among African Americans? *Journal of Acquired Immune Deficiency Syndromes, 38*, 213-218. doi:10.1097/00126334-200502010-00014

Bogart, L. M., & Bird, S. T. (2006). Relationship of African Americans' socio-demographic characteristics to belief in conspiracies about HIV/AIDS and birth control. *Journal of National Medical Association, 98*, 1144. Retrieved from http://www.nmanet.org/

Bowman, B. (2010, February 10). A portrait of Black America on the eve of the 2012 census. *The Root.* Retrieved from http://www.theroot.com

Bristol-Myers Squibb Co. (2000, August 2). Secure the future HIV/AIDS grants distributed among five Sub-Saharan countries. *Biotech Weekly.* Retrieved from http://www.newsrx.com/newsletters/Biotech-Week

Campinha-Bacote, J. (2009). Culture and diversity issues: A culturally competent model of care for African Americans. *Urologic Nursing, 29*(1), 49-54. Retrieved from http://www.suna.org/resources/urologic-nursing-journal

Centers for Disease Control and Prevention. (2005). *Cases of HIV infection and AIDS in the United States and dependent areas, 2005.* Atlanta, GA. Retrieved from http://www.cdc.gov/hiv/surveillance/resources/reports/2005report/

Centers for Disease Control and Prevention. (2006). *2006 STD treatment guidelines.* Atlanta, GA. Retrieved from http://www.cdc.gov/std/treatment/2006/toc.htm

Centers for Disease Control and Prevention. (2007a). *Cases of HIV infection and AIDS in the United States and dependent areas, 2005.* Atlanta, GA. Retrieved from http://www.cdc.gov/hiv/surveillance/resources/reports/2007report/

Centers for Disease Control and Prevention. (2007b, March 9). Racial/ethnic disparities in diagnoses of HIV/AIDS—33 States, 2001–2005. *Morbidity and Mortality Weekly Report, 56*(9), 189-193. Atlanta, GA. Retrieved from http://www.cdc.gov/mmwr/index.html

Centers for Disease Control and Prevention. (2008). *Diagnoses of HIV infection and AIDS in the United States and dependent areas, 2008.* Atlanta, GA. Retrieved from http://www.cdc.gov/hiv/surveillance/resources/reports/2008report

Centers for Disease Control and Prevention. (2009). *HIV among African Americans.* Atlanta, GA. Retrieved from http://www.cdc.gov/hiv/topics

Center for Disease Control and Prevention. (2010). *Disparities in HIV/AIDS, viral hepatitis, STDs and TTB.* Atlanta, GA. Retrieved from http://www.cdc.gov/nchhstp/ healthdisparities/

Centers for Disease Control and Prevention. (2012). *HIV/AIDS.* Atlanta, GA. Retrieved from http://www.cdc.gov/hiv

Communication. (2007). In *Encyclopedia Britannica online* (Academic ed.). Retrieved from http://www.britannica.com/EBchecked/topic/129024/communication

Cornelius, L. J., & Hamilton-Mason, J. (2009). Enduring issues in HIV/AIDS for people of color: What is the road ahead? *Health & Social Work, 34*, 243-247. Retrieved from http://www.naswpress.org/publications/journals/hsw.html

Creswell, J. W. (2008). *Educational research: Planning, conducting, and evaluating quantitative and qualitative research.* Upper Saddle River, NJ: Pearson.

Denzin, N., & Lincoln, Y. (Eds.). (2000). *Handbook of qualitative research.* London, England: Sage.

Erdi, P., Ujfalussy, B., & Diwadkar, V. (2009). The schizophrenic brain: A broken hermeneutic circle. *Neural Network World, 19*, 413-428. Retrieved from http://geza.kzoo.edu/~erdi/bgy09/nnw6.pdf

Ezejiofor, G. C. (2008). Phenomenological study: Role of culture in promoting contraction of HIV/AIDS among Anambra state women, Nigeria. *Dissertation Abstracts International: Section B. Sciences and Engineering, 70*(03). (UMI No. 3349278)

Fenster, M. (2008). *Conspiracy theories: Secrecy and power in American culture.* Minneapolis, MN: University of Minnesota Press.

Foster, J. D., & Cone, J. D. (1998). *Dissertations and theses from start to finish.* Washington, DC: American Psychological Association.

Gandhi, M. (2013) www.UCTV. YouTube.

Gilley, B. J., & Keesee, M. (2007). Linking 'white oppression' and HIV/AIDS in American Indian etiology: Conspiracy beliefs among AI MSMs and their peers. *American Indian and Alaska Native Mental Health Research, 14*(1), 48-66. doi:10.5820/aian.1401.2007.48

Grazer, B. (Producer), & Howard, R. (Director). (1995). *Apollo 13* [Motion picture]. United States: Universal Films.

Greenberg, A. E., Hader, S. L., Masur, H., Young, A. T., Skillicorn, J., & Dieffenbach, C. W. (2009). Fighting HIV/AIDS in Washington, D.C. *Health Aff. (Millwood), 28*, 1677-1687. doi:10.1377/hlthaff.28.6.1677

Hahn, B.W. (2000) AIDS as a Zoonosis: Scientific and Public Health Implications. Science, 2000.

Hyde, J. S., & DeLamater, J. D. (2003). *Understanding human sexuality* (8th ed.). New York, NY: McGraw-Hill.

Johnson, R. (2012, July 19). Save lives: End the HIV stigma. *CNN*. Retrieved from http://www.cnn.com/

King, J. L. (2005). *On the down low: A journey into the lives of straight black men who sleep with men.* New York, NY: Broadway Books.

King, W. D. (2003). Examining African American's mistrust of the healthcare system: Expanding the research question. *Public Health Reports, 118,* 366-367. Retrieved from http://www.publichealthreports.org/

Lazarous, M., Newman, I. M., & Shell, D. F. (2007). Factors contributing to the failure to use condoms among students in Zambia. *Journal of Alcohol and Drug Education, 51*(2), 40-59. Retrieved from http://www.jadejournal.com/

Leedy, P. D., & Ormrod, J. E. (2005). *Practical research: Planning and design* (8th ed.). Upper Saddle River, NJ: Pearson.

Lombardo, P. A., & Dorr, G. M. (2006). Eugenics, medical education, and the public health service: Another perspective on the Tuskegee syphilis experiment. *Bulletin of the History of Medicine, 80,* 291-317. doi:10.1353/bhm.2006.0066

Lucey, B. P., Nelson-Rees, W. A., & Hutchins, G. M. (2009). Henrietta Lacks, HeLa cells and cell culture contamination. *Archives of Pathology Laboratory Medicine, 133*, 1463-1467. doi:10.1043/1543-2165-133.9.1463

Mason, M. (2010). Forum: Sampling size and saturation in PhD studies using qualitative interviews. *Qualitative Social Research, 11*(3), 8. Retrieved from http://www.qualitative-research.net

McCormick, M. J., & Martinko, M. J. (2004). Identifying leader social cognitions: Integrating the causal reasoning perspective into social cognitive theory. *Journal of Leadership and Organizational Studies, 10*, 2-11. doi:10.1177/107179190401000401

McGrath, M. (2009, October 20). HIV vaccine trial was significant. *BBC News*. Retrieved from http://news.bbc.co.uk

Mikesell, R. H., Lusterman, D. D., & McDaniel, S. H. (2003). *Integrating family therapy: Handbook of family psychology and systems theory*. Washington, DC: Sheridan.

Moustakas, C. (1994). *Phenomenological research methods*. Thousand Oaks, CA: Sage.

MSM of color need HIV men's health leadership. (2011). *AIDS Alert, 26*(10), 114-115. Retrieved from http://www.ncbi.nlm.nih.gov/pubmed/22096784

Mulatu, M. S., Leonard, K. J., Godette, D. C., & Fulmore, D. (2008). Disparities in the patterns and determinants of HIV risk behaviors among adolescents entering substance abuse treatment programs. *Journal of the National Medical Association, 100*, 1405-1417. Retrieved from http://www.nmanet.org/

National Association of Social Workers. (2012). *Understanding HIV/AIDS stigma.* Washington, DC. Retrieved from http://www.socialworkers.org/practice/hiv_aids/ AIDS_Day2012.pdf

Neuman, W. L. (2003). *Social research methods* (5th ed.). Upper Saddle River, NJ: Prentice Hall.

N.Y. Times, (2013). The Immortal Life of Henrietta Lacks, the Sequel. http://www.salon.com/2014/09/04/10_of_the_most_evil_medical_experiments_in_history_partner/?utm_source=facebook&utm_medium=socialflow

Peschl, M. F. (2007). Triple-loop learning as foundation for profound change, individual cultivation, and radical innovation: Construction processes beyond scientific and rational knowledge. *Constructivist Foundations, 2*(2-3), 136-145. Retrieved from http://www.univie.ac.at/constructivism/journal/

Peterson, J. L., & Jones, T. K. (2009). HIV prevention for men who have sex with men in the United States. *American Journal of Public Health, 99*, 976-981. doi:10.2105/ AJPH.2008.143214

Primas, E. G. (2008). Faith-based leaders' influence on health issues of African-American women. *Dissertation Abstracts International: Section B. Sciences and Engineering, 69*(08). (UMI No. 3324082)

Schneider, E., Whitmore, S., Glynn, M. K., Dominguez, K., Mitsch, A., & McKenna, M. T. (2008). Revised surveillance case definitions for HIV infection among adults, adolescents, and children aged <18 months and for HIV infection and AIDS among children aged 18 months to <13 years—United States 2008. *Morbidity and Mortality Weekly Report, 57*(RR10), 1-8. Retrieved from http://www.cdc.gov/mmwr/preview/mmwrhtml/ rr5710a1.htm

Siegel, K., Schrimshaw, E. W., & Karus, D. (2004). Racial disparities in sexual risk behaviors and drug use among older gay/bisexual and heterosexual men living with HIV/AIDS. *Journal of the National Medical Association, 96*, 215-224. Retrieved from http://www.nmanet.org/

Simone, R. (Producer/Director). (2012). *End game: AIDS in black America* [Documentary].

Retrieved from http://nova.campusguides.com/content. php?pid= 114919&sid=992788#video

Simoni, J. M., & Ng, M. T. (2002). Trauma, coping, and depression among women with HIV/AIDS in New York City. *AIDS Care, 12*, 567-580. doi:10.1080/ 095401200750003752

Smallman, S. (2008). A case for guarded optimism: HIV/ AIDS in Latin America. *NACLA Report on Americans, 41*(4), 14-19. Retrieved from https://nacla.org/

Stigma. (1994). In *Mosby's pocket dictionary of medicine, nursing, and allied health* (2nd ed.). St. Louis, MO: Mosby.

Sutton, Y. M., Jones, R. L., Wolitski, R. J., Cleveland, J. C., Dean, H. D., & Fenton, K. A. (2009). A review of the Centers for Disease Control and Prevention response to the HIV/AIDS crisis among Blacks in the United States, 1981–2009. *American Journal of Public Health, 99*, S351-S359. doi:10.2105/AJPH.2008.157958

The Peculiar Institution. (2009). *America's Civil War, 22*(4), 36-37. Retrieved from http://www.historynet.com/americas-civil-war

The New York Times. (2013, August 13). *Afraid to get tested? Slow down and think about it.* Retrieved from http://http://well.blogs.nytimes.com

The New York Times. (2012, November 21). *New H.I.V. cases falling in some poor nations, but treatment still lags.* Retrieved from http://health.nytimes.com

Thomas, S. B., & Quinn, S. C. (1991). The Tuskegee syphilis study, 1932 to 1972: Implications for HIV education and AIDS risk education programs in the Black community. *American Journal of Public Health, 81*, 1498-1505. doi:10.2105/AJPH.81.11.1498

Thompson-Robinson, M., Weaver, M., Shegog, M., & Richter, D. (2007). Perceptions of heterosexual African-American males' high-risk sex behaviors. *International Journal of Men's Health, 6*, 156-166. doi:10.3149/jmh.0602.156

Trochim, W. M. K. (2006). *Qualitative validity.* Retrieved from http://www. socialresearchmethods.net/kb/qualval.php

Winfrey, O. (Producer), Perry, T., & Daniels, L. (Directors). (2009). *Precious* [Motion picture]. United States: Lionsgate Films.

Wolfe, N. (2011). The viral storm: The dawn of a new pandemic age. New York, NY: Henry Holt and Company.

World Health Organization. (2006, August 7). *Case definitions of HIV for surveillance and revised clinical staging and immunological classification of HIV related disease in adults and children.* Geneva, Switzerland. Retrieved from http://www. who.int/hiv/pub/guidelines/hivstaging/en/index.html

Wyatt, G. E., Williams, J. K., Henderson, T., & Sumner, K. (2009). On the outside looking in: Promoting HIV/AIDS research initiated by African-American investigators. *American Journal of Public Health, 99,* S48-S53. doi:10.2105/AJPH.2007.131094